T0140776

Bibliografische Information der Deutschen Nationalbibliothek

Die Deutsche Nationalbibliothek verzeichnet diese Publikation in der Deutschen Nationalbibliografie; detaillierte bibliografische Daten sind im Internet über http://dnb.d-nb.de abrufbar.

ISBN 978-3-8325-3323-6

Logos Verlag Berlin GmbH
Comeniushof, Gubener Str. 47,
10243 Berlin
Tel.: +49 (0)30 42 85 10 90
Fax: +49 (0)30 42 85 10 92
INTERNET: http://www.logos-verlag.de

Regularization of Inverse Problems for Turning Processes

von Anna Christina Brandt

Dissertation

zur Erlangung des Grades eines Doktors der Naturwissenschaften
— Dr. rer. nat. —

Vorgelegt im Fachbereich 3 (Mathematik & Informatik)
der Universität Bremen
im September 2012

Datum des Promotionskolloquiums: 19. 12. 2012

Gutachter: Prof. Dr. Peter Maaß (Universität Bremen)
Dr. Oltmann Riemer (Universität Bremen)

Zusammenfassung

In der vorliegenden Arbeit werden inverse Probleme behandelt, die im Zusammenhang von ultrapräzisen Drehprozessen auftreten. Ultrapräzise Drehprozesse werden häufig verwendet, um metallische Werkstücke mit hoher Oberflächenqualität herzustellen, welche beispielsweise in optischen Geräten zum Einsatz kommen. Einen wichtigen Einfluss auf die Oberflächenqualität haben Unwuchten, welche zu Schwingungen der Maschinenstruktur führen und mit dem Zerspanprozess wechselwirken können, was als sogenannte Prozess-Maschinen-Wechselwirkung bezeichnet wird.

Daher wird ein Prozess-Maschinen-Wechselwirkungsmodell entwickelt, mit Hilfe dessen der Einfluss von Unwuchten und Prozessparametern auf die resultierende Oberfläche des bearbeiteten Werkstückes simuliert werden kann. Um die Wechselwirkung berücksichtigen zu können, wird ein neues Mikrokraftmodell für ultrapräzise Drehprozesse erstellt. Das resultierende Wechselwirkungsmodell basiert auf einem nichtlinearen parameterabhängigen System gekoppelter gewöhnlicher Differentialgleichungen. Das zugehörige Vorwärtsproblem wird durch die Abbildung beschrieben, die den Eingangsparametern des Modells die Lösung des Differentialgleichungssystems zuordnet.

Der Großteil der Arbeit umfasst die Inversion des Vorwärtsproblems, d.h. es werden Werkzeugpfade auf der Werkstückoberfläche vorgegeben und optimierte Eingangsparameter wie beispielsweise die Schnitttiefe so bestimmt, dass bei Einsetzen der neuen Parameter in das Vorwärtsproblem die gewünschten vorgegebenen Werkzeugpfade erzeugt werden. Da das Vorwärtsproblem schlecht gestellt ist, werden Regularisierungsmethoden mit "Sparsity-Straftermen" angewandt, welche sparse Lösungen erzeugen. Der Vorteil einer sparsen Lösung liegt darin, dass sie die Anzahl der Steuerpunkte in der Maschinensteuerung begrenzt. Es werden zwei verschiedene Anwendungen, das Designproblem sowie die Berechnung sparser Korrekturen, ausführlich vorgestellt und mit zahlreichen numerischen Beispielen illustriert.

Abstract

In this thesis, we will focus on inverse problems appearing in ultra precise turning processes. Ultra precision turning is widely used to manufacture metallic surfaces with high surface quality which for example can be used for optical devices. One crucial influencing factor of the surface quality is unbalances which lead to vibrations of the machine structure and which can interact with the cutting process. This interaction is the so-called process machine interaction.

Therefore, a process machine interaction model is built which simulates the influence of unbalances of the machine structure and process parameters on the resulting surface of the machined workpiece. In order to include the process machine interaction into the model, a new micro force model for ultra precision turning is developed. The resulting interaction model is mainly based on a coupled, nonlinear parameter-dependent system of ordinary differential equations. The corresponding forward model is thus described by the parameter-to-state map which maps the input parameters to the solution of the differential equation system.

The main part of the thesis is the inversion of the forward operator, i.e. for a given tool path on the workpiece the necessary input parameters like for example the depth of cut are computed such that solving the forward model with this new input parameters results in the desired tool path. Since the forward problem is ill-posed, regularization methods with sparsity constraints are applied which promote sparse solutions. The advantage of such sparse solutions is that they limit the points of machine changes in the machine control. Two different applications, the design problem and the sparse correction problem, are treated in detail and illustrated with various numerical examples.

I would like to thank:

Prof. Dr. Peter Maaß for supervision, support and motivation in writing this thesis.
All the people from the Center for Industrial Mathematics for mathematical and technical support, especially the whole *AG Technomathematik* for the nice working atmosphere and for the fruitful discussions. Dr. Jenny Niebsch, Andreas Krause and Dr. Oltmann Riemer for the rewarding collaboration in the project.
The German Research Foundation for funding my research within the Priority Program SPP 1180.
Moreover, Robin Strehlow, Jost Vehmeyer, Rudolf Ressel, and Fiona Knoll for proofreading parts of this thesis and giving lots of useful remarks.
All the people of my coffee break group for the advices and lively discussions - both mathematically and personally, and especially Dr. Kamil Kazimierski and Dr. Iwona Piotrowska-Kurczewski for their mathematical support but in particular for their friendship and moral encouragement to finish this thesis.
And finally my family and all my friends for their moral support during the whole time.

Christina Brandt
Zentrum für Technomathematik
Universität Bremen

Contents

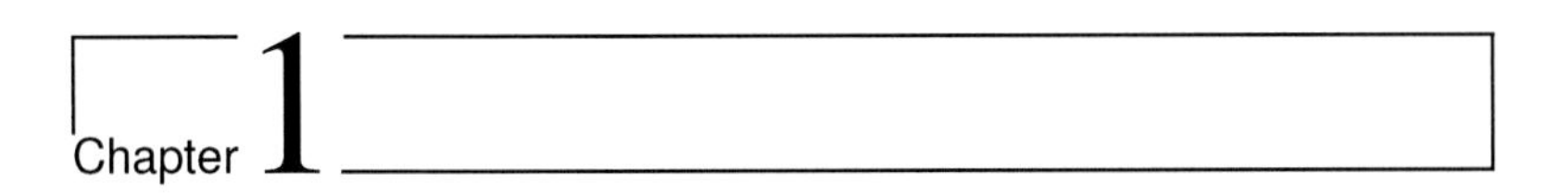

Introduction

In this thesis, we will focus on inverse problems appearing in ultra precise turning processes. Therefore, an introduction about turning as an manufacturing process and inverse problems is given in this chapter.

1.1 Turning processes

Turning is a widely used manufacturing process and belongs to the family of cutting processes like milling, drilling and grinding. It is a metal cutting process where single point tools are applied. Basically, turning is characterized by the rotation of the workpiece and a lateral movement of the tool. It is possible to distinguish between different turning processes regarding to the geometry of the process, e.g. cylindrical turning (tool movement parallel to the rotation axis) for producing cylindrical surfaces, face turning (tool translation perpendicular to the rotation axis) to manufacture planar surfaces or grooving in order to cut an external or internal groove into the workpiece. More information about turning as well as cutting processes in general is given in [1, 59].

Turning is used to manufacture surfaces of metallic workpieces, consisting mostly of rotationally symmetric workpieces. The precision of the process and the quality of the manufactured surfaces has been improved such that the process may be called precision turning or even ultra precision turning. According to Taniguchi [56], precision in this context has a changing meaning with reference to the development of machining processes with increasing accuracy. Currently, ultra precision machining is characterized by relative positioning errors of tool and workpiece less than 10^{-6} [50]. For example, in ultra precision turning the goal is the production of surfaces with a roughness below 10nm and form deviations below 1μm [7]. In order to achieve such a high precision, diamond tools are used with cutting edge sharpened to a radius smaller than 1μm [50].

Therefore, ultra precision turning is applied for manufacturing optical components and mechanical parts with high requirements regarding the surface quality like molds and mirrors, see the examples in Figure 1.1. In industrial applications these optical parts can be used for optical devices like laser systems, scientific instruments, sensors, and many more [49]. Because of the various applications, the development of even more precise machining processes or more complex processes is still an actual research topic. For example, in order to produce diffractive optical elements, a

(a) Mold for pocket camera lens.

(b) Mold for BMW headlight.

(c) Mold for aspherical Fresnel lens.

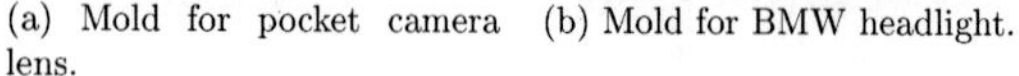

Figure 1.1: Examples of ultra precise turned workpieces, manufactured at the Laboratory of Precision Machining, Bremen.

diamond face turning process with nano fast tool servo has been developed. Hereby, a modulated depth of cut is used, i.e. the depth of cut is changing in the nanometer range with a frequency up to 10 kHz during the process [15].

With increasing precision, more and more disturbing effects have to be understood and minimized. One crucial influencing factor of the surface quality in ultra precision turning are unbalances which lead to vibrations of the machine structure and therefore of the workpiece. Whereas static unbalances can be compensated by balancing the machine before the process [67], there may occur process induced unbalances because the tool itself may cause vibrations while acting on the workpiece. Since residual unbalances of the machine spindle can also influence the process itself, the so-called process machine interaction has become an important research topic in the last years, see [12] and the references therein. The modeling of the process machine interaction in the case of ultra precise turning process is one part of this thesis and is treated in detail in Chapter 3.

1.2 Force models for turning processes

The process machine interaction model is based on the cutting forces of the process and the moments, which they induce to the machine structure. Therefore, the forces have to be modeled accurately. The development of force models for conventional turning processes has been well investigated for decades, and thus many analytical and empiric models can be found in literature. An overview is given for example in [52]. There are linear models like the approach of Weck, where the cutting force

$$F_c = k_c b h$$

is proportional to the product of chip width b and chip thickness h. The proportional factor is the specific cutting force k_c defined as the cutting force divided by the cross sectional area of cut A_c. This model has been improved by Altintas [1, Section 2.5.3] who added a second term in order to incorporate friction caused by the edge forces.

This leads to a force of the form

$$F_c = k_c bh + k_e b\,, \tag{1.1}$$

which is widely applied in many research papers, and which has been unified for different cutting processes with defined cutting edges [32].

On the other hand, there are non-linear models for the cutting force like that one proposed by Kienzle in 1952 [35]. Here, the specific cutting force is represented as

$$k_c = k_{c1,1} h^{-m} \tag{1.2}$$

with the specific cutting force $k_{c_{1,1}}$for a chip width $b_0 = 1\text{mm}$ and a chip thickness $h_0 = 1\text{mm}$. Therefore, the cutting force is given by

$$F_c = k_{c_{1,1}} bh^{1-m}\,, \tag{1.3}$$

where m has to be determined experimentally. The specific cutting force k_c may depend on process parameters, material properties, or process conditions like for example temperature, see [59]. Therefore, there are several approaches for including these effect in the model of the specific cutting force.

Another nonlinear force model for turning with single point tool with nose radius is proposed by Eynian and Altintas, see [21, 20]. The force model is based on the idea that the chip flow is normal to the approximate chord L connecting the tow ends of the cutting edge engaged with the cut. This leads to a force of the form

$$F_c = k_0 + k_L L + k_A A_c\,.$$

Conversely to the situation of conventional cutting processes, the development of force models for micro cutting is an actual research topic because several so-called size effects can occur like the cutting edge radius effect, minimum chip thickness, and ploughing effects. See [61] for an overview about size effects in cutting operations. New force models have been developed like the recently by Jin and Altintas proposed slip-line force model for micro turning with edge tools including strain and temperature effects [29]. The critical chip thickness and micro ploughing effects are also examined, for example by Malekian and Park [41].

In ultra precision turning the situation is exceptional. The cutting parameters like depth of cut and feed rate are in the range of some micrometers which is possible due to the use of diamond tools with much sharper cutting edges than conventional (carbide) tools, and the application of ultra precise machine tools. Therefore, some of the mentioned size effects do play only a subordinate role in diamond cutting. Nevertheless, the forces in the experiments show the typical behavior for micro machining of aluminum alloy (AlMg3), i.e. the passive force is the dominant force component, and the forces are in the range of one Newton or less. Consequently, the mentioned conventional models as well as the micro cutting models are not applicable to ultra precision turning experiments without any additional adaption.

A class of force models, which can easily be adapted for micro turning, are models of Kienzle-type, based on the Kienzle-model (1.3). Several modifications of this model have been developed which are based on different approaches of determining the specific force. In [34] for example, an ansatz is presented by Köhler for the calculation of the undeformed chip thickness h which permits to consider tools with different tool nose geometries. Another modified Kienzle model for micro turning is proposed by Weber et. al. in [62] where the normalized specific cutting force k_c/k_{c0} is represented as a product of functions including cutting velocity v_c, friction μ, uncut chip thickness h, and cutting edge radius r_β, i.e.

$$k_c/k_{c0} = f_1(h)f_2(v_c)f_3(r_\beta)f_4(\mu) \tag{1.4}$$

with functions $f_1(h) = ch^{-m}$, $f_2(v_c) = \alpha_1 v_c^{\beta_1} + \alpha_2 v_c^{-\beta_2}$, $f_3(r_\beta) = a_r + c_r r_\beta$, and $f_4(\mu) = a_\mu + c_\mu \mu$. With k_{c0} the specific cutting force for standard conditions is denoted. Because of the normalization, each function f_i equals zero in the case of standard cutting conditions, and consequently the advantage of the force model is that the model parameters of each function f_i can be determined separately. This force model is used in Chapter 3 as starting point for the development of a new force model for ultra precision turning.

1.3 Inverse problems for turning operations

The second part of this thesis concerns the regularization of inverse problems appearing in turning operations. In particular, the computation of optimized input parameters for the turning process is a typical inverse problem. Solving an inverse problem means the inversion of the direct problem. The direct problem is mathematically formulated as an operator equation of the form

$$F(x) = y\,,$$

where the operator F describes the model, x contains the relevant model parameter, and y is the resulting output of the model. Thinking of the parameter identification problem of the turning process, x may be thought as the depth of cut of the tool. The operator F is a model which computes the resulting surface y of the turning process, when x is used as input of the process. Solving the direct problem would mean to compute the surface y for given input x and known process model F.

Assume now that a desired surface is given. The inverse problem is then the task of determining the unknown input x for given surface y and known process model F such that applying the computed input would result to the desired surface y. Note that in general the unknown parameter may be a function of time or spatial variable.

Inverse problems are usually so-called ill-posed problems. Even if an inverse operator F^{-1} is known, it can be discontinuous. Therefore, applying F^{-1} to noisy measurements y of surfaces would lead to bad reconstructions of x. That is why regularization methods are necessary in order to stabilize the inversion. The main idea behind these methods is to compute a sequence of approximations of F^{-1} which are continuous and which will converge for decreasing noise level to the real inverse operator F^{-1}. Or to say it in a simple way, instead of solving the correct problem badly, we like to solve the bad problem correctly. A short introduction into inverse problems and regularization methods will be given in Chapter 2.

In the last years, the concept of sparsity and regularization with sparsity constraints have become more and more important. Sparsity means in this context that a mathematical variable or function x should be expressed in a certain basis (or frame) of basis vectors $\{\varphi_i\}_{i\in\mathbb{N}}$ with as few as possible basis vectors. Formally, the function x can be expanded as the sum

$$x = \sum_{i=1}^{\infty} x_i \varphi_i \,,$$

and it is called sparse in the basis $\{\varphi_i\}_{i\in\mathbb{N}}$ if a finite number of coefficients x_i are non-zero. Coming back again to the example of computing the depth of cut as input parameter of the process model, it may be convenient to compute a solution which is sparse in the set of piecewise constant functions. The advantage of such a solution is that it has a few necessary points of machine changes in the machine control. In contrast, determining x without the additional assumption of a sparse solution may lead to infinitely many coefficients x_i and consequently to an input of the machine for which the machine control has to change the positions at very fine time steps. The concept of sparsity can be found in various applications, e.g denoising of signals, image processing, or tomography.

The computation of sparse input parameters for turning processes is the main topic of this thesis. Therefore, a model for the turning process is developed which includes cutting forces, tool and workpiece vibrations as well as process machine interaction, and which computes the tool path on the workpiece surface and therefore the resulting surface itself. The underlying model is based on a nonlinear parameter-dependent system of coupled ordinary differential equations. The forward operator is thus the operator mapping the input parameter to the solution of the differential equation system. Regularization methods with sparsity constraints will be applied in order to compute sparse input parameters for the cutting process.

1.4 Overview about the thesis

The thesis is organized as follows. The mathematical basics for this thesis are given in Chapter 2. First, mathematical definitions and notations are introduced in Section 2.1, and basic concepts of convex analysis are explained in Section 2.2. The solution theory of parameter-dependent ordinary differential equations is presented in Section 2.3, whereas in Section 2.4 an introduction to the theory of inverse problems and their regularization is given.

In Chapter 3 the underlying forward model of the turning process is developed. The main part is the process machine interaction model which is built in several steps in Section 3.2. It includes the process model describing the process forces and process parameters (Subsection 3.2.1), the machine model that computes the machine vibrations for given unbalances (Subsection 3.2.2) and finally the coupling of both sub-models (Subsection 3.2.3). In Section 3.3, it is shown how the resulting surface is simulated using the output of the process machine interaction model.

The process machine interaction model is the basis for Chapter 4, where the parameter identification problem for turning processes is solved. First, the forward operator is formulated and analyzed in Section 4.1. Then the identification problem is treated in the following section. Therefore, the problem is reformulated such that two linear operator equations have to be solved (Subsection 4.2.2). Different applications like the design problem and the sparse correction problem are presented and illustrated with various numerical examples in Subsection 4.2.

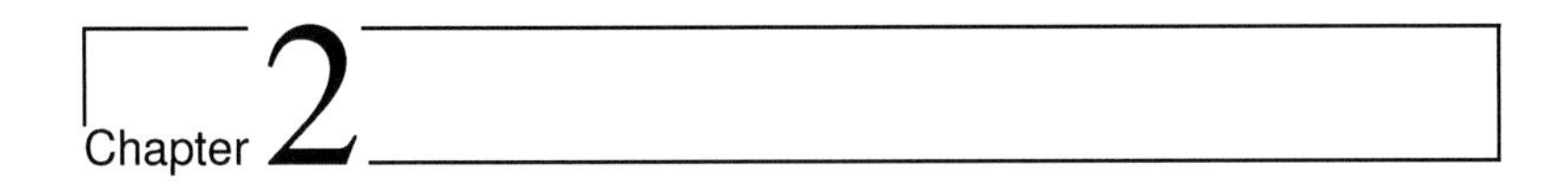

Chapter 2

Preliminaries and concepts

This chapter provides an overview about the mathematical concepts and methods needed in this thesis. First, basic notations and definitions are introduced in Section 2.1, and an introduction to convex analysis is given in Section 2.2. In the following Section 2.3, an overview about the theory of parameter-dependent ordinary differential equations is presented which is the basis for the analysis of the forward problem in Chapter 4. Finally, this chapter concludes with the theory of inverse problems and regularization, which is the main mathematical topic of this thesis.

2.1 Basics: notations and definitions

This section concerns basic notations and definitions needed in the following. As usual we denote with $C(D, E)$ the space of continuous functions between Banach spaces D and E. An additional exponent $k \in \mathbb{N}$ indicates the subspace $C^k(D, E)$ of k-times continuously differential maps from $D \subset E$ open in E, whereas an index 0 stands for the subset of functions with compact support.

We assume that the reader is familiar with the basics of measure and integration theory. In the following a collection of basic notations needed in this thesis are briefly presented. For more information, see standard textbooks about measure theory, e.g. [30].

Let $(\Omega, \mathcal{A}, \mu)$ be a measure space with measure $\mu : \mathcal{A} \to \mathbb{R}_+ \cup \{\infty\}$ on a σ-algebra $\mathcal{A}$ of measurable subsets of the nonempty set Ω. If μ is the Lebesgue measure, we will call the function measurable instead of Lebesgue-measurable. As usually, a measurable subset A of Ω is called (μ-)*null set* if $\mu(A) = 0$, and we say that a property holds *almost everywhere* (a.e.) in Ω or for *almost all* (a.a.) $t \in \Omega$, if there is a null set A such that the property holds for all $t \in \Omega \setminus A$. A function $f : \Omega \to E$ mapping into a Banach space E is called *μ-measurable* if for each open set $U \subset E$, $f^{-1}(U)$ is μ-measurable.

By a simple function or step function we mean a function $s : \Omega \to E$ which assumes only a finite number of distinct values, i.e. which can be represented as a sum of piecewise constant functions $s = \sum_{i=1}^{n} a_i \chi_{A_i}$ where the sets A_i are disjoint and μ-measurable and the elements $a_i \in E$ are distinct. Hereby, χ_A denotes the characteristic function of A. A μ-measurable function $f : \Omega \to E$ can be represented as the limit of a sequence $\{s_i\}_i$ of μ-measurable simple functions.

As usual, the representation of a measurable simple function as a sum of piecewise constant functions can be used to define the (Lebesgue-) integral, i.e.

$$\int_{\Omega} s \, \mathrm{d}\mu = \sum_{i=1}^{n} a_i \mu(A_i) .$$

This definition of the integral of a step function can be used to define the integral of a measurable function.

Definition 1. Let $f : \Omega \to E$ be μ-measurable and $\{s_n\}_n$ an approximating sequence of simple functions, i.e. $\lim_{n\to\infty} s_n(t) = f(t)$ for almost all $t \in \Omega$ and $\int_{\Omega} \|s_n - s_m\| \, \mathrm{d}\mu \to 0$ as $n, m \to \infty$. Then the function f is said to be *μ-integrable.* The integral is defined by

$$\int_{\Omega} f \, \mathrm{d}\mu := \lim_{n\to\infty} \int_{\Omega} s_n \, \mathrm{d}\mu .$$

The collection of all integrable functions will be denoted by $L^1(\Omega, E)$. The space $L^1(\Omega, E)$ is a Banach space with norm $\|f\|_1 = \int_{\Omega} \|f(t)\| \, \mathrm{d}\mu$, where $\|f\|_1 = 0$, if and only if $f(t) = 0$ for a.a. $t \in \Omega$.

As usual, a μ-measurable function $f : \Omega \to E$ is said to be p-integrable for $p \in \mathbb{N}$ if

$$\int_{\Omega} \|f\|_E^p \, \mathrm{d}\mu < \infty .$$

The space $L^p(\Omega, E)$ of p-integrable functions is then defined as

$$L^p(\Omega, E) = \left\{ f : \Omega \to E \,|\, f \ \mu\text{-measurable}, \ \|f\|_p := \left(\int_{\Omega} \|f\|_E^p \, \mathrm{d}\mu \right)^{1/p} < \infty \right\} .$$

Definition 2. Let $f : \Omega \to P$ be a measurable function into a metric space P. Then f is said to be *essentially bounded* if there exists a compact subset $K \subseteq P$ such that

$$f(t) \in K \qquad \text{for almost all } t \in \Omega .$$

It is called *locally essentially bounded* if the restriction of f to every bounded subset of Ω is essentially bounded.

With this definition it is possible to define the space of all essentially bounded functions.

Definition 3. Denote the set of all measurable and essentially bounded functions $f : \mathcal{I} \to P$ with $L^{\infty}(\mathcal{I}, P)$ and the set of all measurable and locally essentially bounded functions with $L^{\infty}_{\mathrm{loc}}(\mathcal{I}, P)$. Moreover, the norm of a function $f \in L^{\infty}(\mathcal{I}, P)$ is given by

$$\|f\|_{\infty} := \operatorname*{ess.\,sup}_{t \in \mathcal{I}} \{\|f(t)\|_E\} = \inf\{c \in \mathbb{R} \,|\, \{\|f(t)\| > c\} \text{ has measure zero}\} .$$

Next Sobolev spaces and the concept of weak differentiability are introduced.

Definition 4. Let $f \in L^1(\mathcal{I}, E)$. It there exits a function $g \in L^1(\mathcal{I}, E)$ such that

$$\int_{\mathcal{I}} \varphi' f \, \mathrm{d}\mu = -\int_{\mathcal{I}} \varphi g \, \mathrm{d}\mu, \qquad \text{for all } \varphi \in C_0^{\infty}(\mathcal{I}) ,$$

then f is *weakly differentiable,* and $g = f'$ is called the *weak derivative* of f .

The space $W^{1,p}(\mathcal{I},E)$ is defined as

$$W^{1,p}(\mathcal{I},E) := \{f:\mathcal{I}\to E \,|\, f\in L^p(\mathcal{I},E),\, f'\in L^p(\mathcal{I},E)\}.$$

It can be equipped with the norm

$$\|f\|_{W^{1,p}} := \begin{cases} \int_{\mathcal{I}} (\|f\|_E^p + \|f'\|^p)^{1/p}, & 1\le p<\infty \\ \text{ess.}\sup\{\|f(t)\|_E + \|f'(t)\|_E,\, t\in\mathcal{I}\}, & p=\infty. \end{cases}$$

The next definition will be used in order to introduce a generalized solution concept for solutions of differential equation in Section 2.3.

Definition 5. Let E be a Banach space. The function $u:\mathcal{I}\to E$ defined on a compact interval $\mathcal{I}=[a,b]$, $a<b$, is *absolutely continuous,* if for every $\epsilon>0$, there exists $\delta>0$ such that for every k and every sequence of points

$$a\le a_1<b_1\le a_2<b_2\le\ldots a_k<b_k\le b$$

with

$$\sum_{i=1}^{k}(b_i-a_i)<\delta,$$

it holds that

$$\sum_{i=1}^{k}|u(b_i)-u(a_i)|<\epsilon.$$

We will denote by $AC(\mathcal{I},E)$ the space of all absolutely continuous functions over the interval $\mathcal{I}$ into E. Note that an absolutely continuous function is of strong bounded variation, and each absolutely continuous function with essentially bounded derivative is Lipschitz continuous. The following theorem provides an useful characterization of absolute continuity.

Theorem 6. *Suppose that $u:\mathcal{I}\to\mathbb{R}^n$ with $\mathcal{I}=[a,b]$. Then u is absolutely continuous if and only if there exists a function $h\in L^1(\mathcal{I},\mathbb{R}^n)$ such that*

$$u(t)=u(a)+\int_a^t h(\tau)\,d\tau \;(t\in\mathcal{I}). \tag{2.1}$$

It follows that an absolutely continuous function is differentiable for almost all $t\in\mathcal{I}$, i.e. $u'=h$ almost everywhere.

Proof. The proof can be found in [30, Chapter 16]. □

Remark 7. The theorem is valid also for functions $u:\mathcal{I}\to E$, *if* E *is* a n-dimensional or a reflexive Banach space but not for every arbitrary Banach space, like the following example illustrates (see [26, Section 1.4.7]). Consider the space $E=(c_0)=\{x=\{x_n\}_n \,|\, \lim_n x_n=0\}$ with norm $\|x\|_{c_0}=\max_n|x_n|$ and $\mathcal{I}=[0,2\pi]$. Take the function $u:\mathcal{I}\to(c_0)$, $u(t):=\left\{\frac{1}{n}sin(nt)\right\}_n$, $t\in\mathcal{I}$. Then u is absolutely continuous because

$$\|u(t)-u(s)\|_{c_0}=\max_n\left|\frac{1}{n}sin(nt)-\frac{1}{n}sin(ns)\right|\le|t-s|,\; s,t\in\mathcal{I}.$$

If u is differentiable at $t\in\mathcal{I}$, then $u'(t)=\{\cos(nt)\}_n$, but $\lim_n\cos(nt)\neq0$ for all $t\in\mathcal{I}$. Therefore u' does not belong to (c_0), and u can not be absolutely continuous (compare with Theorem 1.4.6 in [26]).

As a consequence of Theorem 6, the space of absolutely continuous functions can be identified with the Sobolev space $W^{1,1}([a,b],\mathbb{R}^n)$.

2.2 Convex Analysis

In this section some notations and basic facts about convex analysis needed in the following are introduced. For a deeper discussion of convex analysis in Banach spaces, we refer to [16] or [51].

In order to minimize functionals in the following, derivatives of functionals are needed and introduced in the next definition. They can be regarded as an analogous case to directional and total derivatives of real functions.

Definition 8 (F- and G-differentiable)**.** Let X and Y Bananch spaces and $f : \mathcal{U} \to Y$ a given map from $\mathcal{U} \subseteq X$ open to Y.

1. The map f is called *Gâteaux-differentiable* (G-differentiable) at $x \in \mathcal{U}$ if there exits a linear map $T \in L(X, Y)$ such that

$$f(x + th) - f(x) - tT(h) = o(t) \;, \qquad t \to 0$$

for all h with $\|h\| = 1$. The map $T =: f'(x)$ is the *Gâteaux-derivative* of f at x. The *Gâteaux-differential* at x is defined by $\delta_G f(x, h) = f'(x)h$.

2. The map f is called *Fréchet-differentiable* (F-differentiable) at $x \in \mathcal{U}$ if there exits a linear map $T \in L(X, Y)$ such that

$$f(x + h) - f(x) - Th = R(x, h)$$

with

$$\lim_{\|h\| \to 0} \frac{\|R(x_0, h)\|}{\|h\|} = 0$$

for all h is some neighborhood of zero. If it exits, $T =: f'(x)$ is called *F-derivative at* x, and the F-differential at x is defined by $df(x; h) = f'(x)h$.

3. If the Fréchet-derivative (resp. Gâteaux-derivative) exits for all $x \in \mathcal{U}$, then the mapping

$$f' : \mathcal{U} \subseteq X \to L(X, Y) \qquad x \mapsto f'(x)$$

is called Fréchet-derivative (resp. Gâteaux-derivative) of f on $\mathcal{U}$.

The next example illustrates the computation of the Gâteaux-derivative of the so-called discrepancy term of a Tikhonov-functional which will be introduced in Subsection 2.4.1.

Example 9 (Calculation of the Gâteaux-derivative of the discrepancy term)**.** Let $K : H_1 \to H$ being a linear operator between Hilbert spaces: The Gâteaux-derivative of $\frac{1}{2} \|K(\cdot) - y\|_H^2 : H_1 \to \mathbb{R}$ should be computed. Therefore, set $\Psi(x) := \frac{1}{2} \|Kx - y\|_H^2$ and compute

$$\begin{aligned} \Psi(x + th) &= \frac{1}{2} \|(Kx - y) + tKh\|_H^2 , \\ &= \frac{1}{2} \left(\|Kx - y\|_H^2 + t^2 \|Kh\|_H^2 + 2t\mathrm{Re} \langle Kx - y \,|Kh \rangle_H \right) , \\ &= \frac{1}{2}\Psi(x) + \frac{1}{2} t^2 \|Kh\|_H^2 + t\mathrm{Re} \langle K^* (Kx - y) \,|h \rangle_{H_1} . \end{aligned}$$

Thus, if follows that

$$\begin{aligned}\frac{1}{t}\left(\Psi(x+th)-\Psi(x)\right) &= \frac{1}{2}t\left\|Kh\right\|_H^2+\operatorname{Re}\left\langle K^*\left(Kx-y\right)|h\right\rangle_{H_1}, \\ &\underset{t\to 0}{\longrightarrow} \operatorname{Re}\left\langle K^*\left(Kx-y\right)|h\right\rangle_{H_1}.\end{aligned}$$

According to Definition 8, the Gâteaux-differential is given by

$$\delta_G\Psi\left(x,h\right)=\operatorname{Re}\left\langle K^*\left(Kx-y\right)|h\right\rangle_{H_1},$$

and therefore the Gâteaux-derivative is $\Psi'(x)=\operatorname{Re}\left\langle K^*\left(Kx-y\right)|(\cdot)\right\rangle_{H_1}$.

In the same way, it is possible to define partial derivatives of functionals mapping between Banach spaces.

Definition 10. Let there be given a map $f:\mathcal{D}(f)\subseteq X\times Y\to Z$ by $(x,y)\mapsto f(x,y)$, where X,Y and Z are Banach spaces. For fixed y set $g(x)=f(x,y)$. If g is F-differentiable (resp. G-differentiable) at x, the partial $F-$ *derivative* (resp. G-derivative) of f with respect to the first variable x is defined as

$$f_x(x,y)=g'(x).$$

Similarly the derivative $f_y(x,y)$ with respect to y is defined. We denote $f_x(x,y)$ and $f_y(x,y)$ also with $D_1f(x,y)$ and $D_2f(x,y)$.

For functionals which are not differentiable in the sense of Gâteaux or Fréchet the term of a derivative can be further generalized by introducing the so-called subdifferential. Since in the following the subdifferential is defined for convex functionals, first the term of convexity is introduced.

Definition 11. Let X be a real space. The function $f:X\to\mathbb{R}\cup\{-\infty,\infty\}$ is called *convex* if for $x,y\in X$ and $\lambda\in[0,1]$ holds that

$$f\left(\lambda x+(1-\lambda)y\right)\leq\lambda f\left(x\right)+(1-\lambda)f\left(y\right).$$

The function is said to be *strictly convex*, if the above inequality is valid with $<$ instead of $\leq$ for all $\lambda\in(0,1)$ and for $x,y\in X$ with $x\neq y$. The *domain* $\mathcal{D}\left(f\right)$ of a convex function is given by $\mathcal{D}(f)=\{x\in X|\,f(x)<\infty\}$.

Definition 12 (Subgradient of a complex function)**.** Let X be a real Banach space and $f:X\to\mathbb{R}\cup\{-\infty,\infty\}$ a convex function. An element $x^*\in X^*$ is a *subgradient* of f in x_0 if

$$f\left(x\right)\geq f\left(x_0\right)+\left(x^*,x-x_0\right)\quad\text{for all } x\in X$$

and $f\left(x_0\right)\neq\pm\infty$. The *subdifferential* of f in x_0 is defined as the set of all subgradients of f in x_0 and is denoted with $\partial f\left(x_0\right)$.

Subgradients are a generalization of the classical derivative. If the Gâteaux-derivative $f'(x_0)$ exits, then the subdifferential is single-valued and given by $\partial f(x_0)=\{f'(x_0)\}$. In the inverse direction, if f is continuous and $\partial f(x_0)$ comprises exactly one element, then the Gateaux-derivative exits and the equality holds again. For a proof of this fact, see [65, Proposition 47.13].

Note that a subdifferential of a single-valued functional may be a set-valued mapping, see the following example. For more information of the concepts of set-valued mappings, see for example [51].

Example 13. Consider the function $f : \mathbb{R} \to \mathbb{R}$ with $f(x) = |x|$. The subdifferential of f is given by the set-valued sign-function $\operatorname{sgn}(x)$:

$$\partial f(x) = \operatorname{sgn}(x) := \begin{cases} -1\,, & x < 0\,, \\ [-1,1]\,, & x = 0\,, \\ 1\,, & x > 0\,. \end{cases}$$

In the classical case, the derivative at $x = 0$ would not exit. Geometrically, the subgradient at this point is the set of all tangents with slopes between -1 and 1, see Figure 2.1.

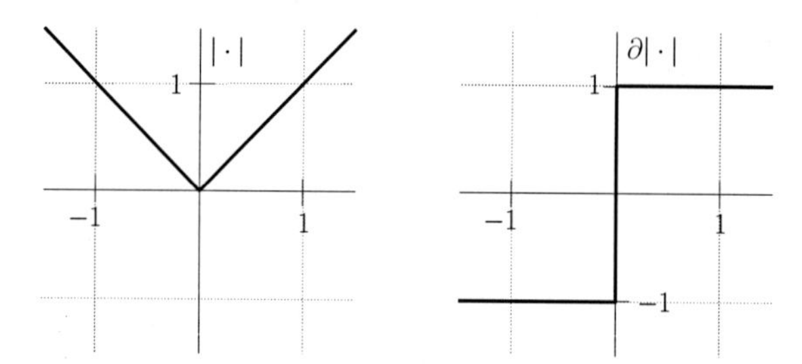

Figure 2.1: The absolute value and the sign-function as its subdifferential. The subgradient at 0 is set-valued. Geometrically, it can be seen that the subgradient at 0 is the set of all possible tangents through $(0, 0)$, with slopes between -1 and 1 which lie below the curve of $|\cdot|$.

In analogy with the classical derivative, there are several important rules how to calculate with subgradients. The most important ones are summarized in the following propositions. Proofs can be found for example in [65].

Proposition 14. *Let X be a real Banach space, $f : X \to \mathbb{R} \cup \{-\infty, \infty\}$ a convex function and $\lambda > 0$. Then it holds*

$$\partial f(\lambda x) = \lambda \partial f(x)$$

for all $x \in X$.

Proposition 15. *[Moreau and Rockafellar] Let X be a real Banach space and $f, g : X \to \mathbb{R} \cup \{-\infty, \infty\}$ convex functions. Assume that there is a point $x_0 \in \mathcal{D}(f) \cap \mathcal{D}(g)$, where f is continuous. Then it holds*

$$\partial (f + g)(x) = \partial f(x) + \partial g(x)$$

for all $x \in X$.

Proof. See [65], Theorem 47.B. □

The next proposition provides a statement about convex minimization problems in real Banach spaces and its connection to the subgradient.

Proposition 16. *Let X be a real Banach space and $f : X \to \mathbb{R} \cup \{\infty\}$ a convex and proper functional, i.e. $f \neq \infty$. Under this assumptions, $x \in X$ is a solution of*

$$\min_{x \in X} f(x) = \alpha$$

if and only if the Euler equation

$$0 \in \partial f(x) \tag{2.2}$$

holds.

Proof. See [65], Proposition 47.12. □

In Subsection 2.4.4 the tools of convex analysis will be applied for the minimization of functionals with sparsity constraints.

2.3 Review about ordinary differential equations

In this section the theory of parameter-dependent ordinary differential equations (ODE) is presented, which will be used in Section 4.1 in order to analyze the forward operator mapping the input parameter to the solution of an ODE. The classical theory of ODEs is treated in standard text books; a nice description of ODEs in n-dimensional Banach-spaces can be found in [2]. It contains also the case of ODEs with constant parameter. For the case of infinitely dimensional Banach-spaces we refer to [18].

However, the theory for the case of time-dependent parameters is not covered in most of the standard books about ODEs. Therefore, the basic facts like existence of a solution and its continuous dependency on the parameter are collected in this section. Since the parameter may be a discontinuous function, the right-hand side of the ODE can be discontinuous which leads to the theory of Carathéodory differential equations [22]. There is also a strong connection to control theory where the parameter plays the role of the (discontinuous) control function of the ODE, see e.g. [55]. The remainder of this sections will mainly follow the lines of [2, 55] and combines therefore the solution theory in Banach spaces and the Carathéodory theory.

2.3.1 Existence theory for initial value problems

In this section, the existence theory of ordinary differential equations without parameter dependence is treated. This is the foundation for the parameter-dependent case in the following subsection. Before starting to proof the existence of solutions, some auxiliary inequalities are derived like the Lemma of Gronwall well-known for the classical theory of ODE.

In the following, let $\mathcal{I} \subseteq \mathbb{R}$ an open interval, $E = (E, |\cdot|)$ be a n-dimensional Banach space over $\mathbb{K}$ and $D \subseteq E$ be open. Moreover, let $f : \mathcal{I} \times D \to E$ be a Carathéodory function, i.e. a function satisfying the two properties that

$$f(.,\xi) : \mathcal{I} \to E \quad \text{is measurable for each fixed } \xi \tag{F1}$$

and

$$f(\sigma,.) : D \to E \quad \text{is continuous for each fixed } \sigma. \tag{F2}$$

These two properties guarantee that $f(t, u(t))$ is measurable as a function of $t \in \mathcal{I}$, see Theorem 62 in the appendix for a proof.

We consider the following initial value problem (IVP)

$$\dot{u}(t) = f(t, u(t)), \qquad u(t_0) = \xi_0 \,. \tag{2.3}$$

Since we do not assume the function f to be continuous in t, the solution u may not be differentiable. This is the motivation of the following definition of the solution of the IVP.

Definition 17. A *solution of the initial value problem* (2.3) on the interval $J \subseteq \mathcal{I}$ containing t_0 is an absolutely continuous function $u : J \to D$ such that

$$u(t) = u_0 + \int_{t_0}^{t} f(\tau, u(\tau))\, \mathrm{d}\tau \tag{INT}$$

holds for all $t \in J$.

Our aim in the following is to prove the existence of such a solution of the IVP (2.3) under certain conditions and to show continuous dependency of the solution from initial conditions. We begin proving a general inequality needed to establish Lemma 20 which is crucial for the existence theorem.

Lemma 18 (Lemma of Gronwall, [2], Lemma 6.1)**.** *Let $\mathcal{I}$ be an interval in $\mathbb{R}$, $t_0 \in \mathcal{I}$ and take functions*

$$\alpha, \beta, \mu : \mathcal{I} \to \mathbb{R}_+$$

such that α and β are locally integrable and μ continuous. If μ satisfies the inequality

$$\mu(t) \le \beta(t) + \left| \int_{t_0}^{t} \alpha(s)\mu(s)\, \mathrm{d}s \right| \quad \forall t \in \mathcal{I}\,, \tag{2.4}$$

then it holds that

$$\mu(t) \le \beta(t) + \left| \int_{t_0}^{t} \alpha(s)\beta(s) e^{\left|\int_s^t \alpha(\tau)\, d\tau\right|} \mathrm{d}s \right| \quad \forall t \in \mathcal{I}\,. \tag{2.5}$$

Proof. Define $v(t) := \int_{t_0}^{t} \alpha(s)\mu(s)\mathrm{d}s$ and observe that it follows from (2.4) that

$$\dot{v}(t) = \alpha(t)\mu(t) \le \alpha(t)\beta(t) + \operatorname{sgn}(t - t_0)\, \alpha(t) v(t)\,, \quad t \in \mathcal{I}\,.$$

Multiplying this inequality with

$$\pi(t) = \mathrm{e}^{-\left|\int_{t_0}^{t} \alpha(s)\, \mathrm{d}s\right|} = \mathrm{e}^{-\int_{t_0}^{t} \operatorname{sgn}(s-t_0)\alpha(s)\, \mathrm{d}s}$$

yields

$$\pi \dot{v} \le \alpha\beta\pi - v\dot{\pi}\,,$$

what can be rewritten as

$$\dot{(\pi v)} \le \alpha\beta\pi\,.$$

By Integration of the inequality, we obtain

$$\begin{aligned} \operatorname{sgn}(t-t_0)v(t)\pi(t) &\leq \operatorname{sgn}(t-t_0)\int_{t_0}^{t}\alpha(s)\beta(s)\pi(s)\,\mathrm{d}s\,, \\ &= \left|\int_{t_0}^{t}\alpha(s)\beta(s)\pi(s)\,\mathrm{d}s\right|, \quad t\in\mathcal{I}\,. \end{aligned}$$

Dividing by $\pi(t)$ we can rewrite the inequality as

$$\operatorname{sgn}(t-t_0)v(t) \leq \left|\int_{t_0}^{t}\alpha(s)\beta(s)\pi(s)/\pi(t)\,\mathrm{d}s\right|.$$

Using (2.4) we can conclude that

$$\begin{aligned} \mu(t) &\leq \beta(t)+\left|\int_{t_0}^{t}\alpha(s)\mu(s)\,\mathrm{d}s\right|, \\ &= \beta(t)+\operatorname{sgn}(t-t_0)v(t)\,, \\ &\leq \beta(t)+\left|\int_{t_0}^{t}\alpha(s)\beta(s)\pi(s)/\pi(t)\,\mathrm{d}s\right|, \\ &= \beta(t)+\left|\int_{t_0}^{t}\alpha(s)\beta(s)\mathrm{e}^{\left|\int_s^t\alpha(\tau)\,\mathrm{d}\tau\right|}\,\mathrm{d}s\right|, \qquad t\in\mathcal{I}\,, \end{aligned}$$

which establishes the desired estimation. □

Corollary 19. *Assume that $\beta(t)=\beta_0\left(|t-t_0|\right)$ with $\beta_0\in C\left(\mathbb{R}_+,\mathbb{R}_+\right)$ is a monotone increasing function and that the inequality*

$$\mu(t)\leq\beta(t)+\left|\int_{t_0}^{t}\alpha(s)\mu(s)\,\mathrm{d}s\right| \quad \forall t\in\mathcal{I} \tag{2.6}$$

holds. Then it follows that

$$\mu(t)\leq\beta(t)\mathrm{e}^{\left|\int_{t_0}^t\alpha(s)\,\mathrm{d}s\right|} \quad \forall t\in\mathcal{I}\,. \tag{2.7}$$

Proof. By the assumptions for β it follows $\beta(s)\leq\beta(t)$ for $|s-t_0|\leq|t-t_0|$, and hence the inequality (2.5) of the previous lemma leads to

$$\begin{aligned} \mu(t) &\leq \beta(t)+\left|\int_{t_0}^{t}\alpha(s)\beta(s)\mathrm{e}^{\left|\int_s^t\alpha(\tau)\,\mathrm{d}\tau\right|}\,\mathrm{d}s\right|, \\ &\leq \beta(t)\left[1+\left|\int_{t_0}^{t}\alpha(s)\mathrm{e}^{\left|\int_s^t\alpha(\tau)\,\mathrm{d}\tau\right|}\,\mathrm{d}s\right|\right], \\ &= \beta(t)\left[1+\operatorname{sgn}\left(t-t_0\right)\int_{t_0}^{t}\alpha(s)\mathrm{e}^{\operatorname{sgn}(t-t_0)\int_s^t\alpha(\tau)\,\mathrm{d}\tau}\,\mathrm{d}s\right], \\ &= \beta(t)\left[1-\left.\mathrm{e}^{\operatorname{sgn}(t-t_0)\int_s^t\alpha(\tau)\,\mathrm{d}\tau}\right|_{s=t_0}^{t}\right], \\ &= \beta(t)\mathrm{e}^{\left|\int_{t_0}^t\alpha(\tau)\,\mathrm{d}\tau\right|}, \end{aligned}$$

and (2.7) is proved. □

The following lemma is simple but the stated inequality will be used repeatedly in the following proofs.

Lemma 20. *Given an interval $\mathcal{I} \subseteq \mathbb{R}$ and two subsets $X_0 \subseteq X \subseteq E$, assume that two functions $f, g : \mathcal{I} \times D \to E$ both satisfy the assumptions* (F1) *and* (F2) *of this section. Suppose furthermore that the two functions*

$$u : \mathcal{I} \to X \quad \textit{and} \quad v : \mathcal{I} \to X_0$$

are continuous. Moreover, assume that there exist two locally integrable functions

$$\alpha, \beta \in \mathrm{L}_{\mathrm{loc}}\left(\mathcal{I}, \mathbb{R}_+\right)$$

such that

$$\left\| f\left(\tau, \xi\right) - f\left(\tau, \eta\right) \right\| \leq \alpha(\tau) \left\| \xi - \eta \right\| \quad \forall \xi, \eta \in X\,, \forall \tau \in \mathcal{I}\,, \tag{2.8}$$

and

$$\left\| f\left(\tau, \xi\right) - g\left(\tau, \xi\right) \right\| \leq \beta(\tau) \quad \forall \xi \in X_0\,, \forall \tau \in \mathcal{I}\,. \tag{2.9}$$

If $t_0 \in \mathcal{I}$ and ξ_0 and η_0 are arbitrary elements of E, then it holds for

$$\tilde{u}(t) := \xi_0 + \int_{t_0}^{t} f\left(\tau, u(\tau)\right)\, d\tau$$

and

$$\tilde{v}(t) := \eta_0 + \int_{t_0}^{t} g\left(\tau, v(\tau)\right)\, d\tau$$

that

$$\left\| \tilde{u}(t) - \tilde{v}(t) \right\| \leq \left\| \xi_0 - \eta_0 \right\| + \int_{t_0}^{t} \alpha(\tau) \left\| u(\tau) - v(\tau) \right\|\, d\tau + \int_{t_0}^{t} \beta(\tau)\, d\tau \tag{2.10}$$

for all $t > t_0$, $t \in \mathcal{I}$.

Proof. Using the triangle inequality and the assumptions (2.8) and (2.9), we deduce that

$$\begin{aligned} \left\| f\left(\tau, \xi\right) - g\left(\tau, \eta\right) \right\| &\leq \left\| f\left(\tau, \xi\right) - f\left(\tau, \eta\right) \right\| + \left\| f\left(\tau, \eta\right) - g\left(\tau, \eta\right) \right\|\,, \\ &\leq \alpha(\tau) \left\| \xi - \eta \right\| + \beta(\tau) \quad \forall \xi \in D,\ \forall \eta \in D_0,\ \forall \tau \in \mathcal{I}\,. \end{aligned}$$

With help of this inequality and again by the triangle inequality we can estimate

$$\begin{aligned} \left\| \tilde{u}(t) - \tilde{v}(t) \right\| &\leq \left\| \xi_0 - \eta_0 \right\| + \left\| \int_{t_0}^{t} f\left(\tau, u(\tau)\right) - g\left(\tau, v(\tau)\right)\, \mathrm{d}\tau \right\|, \\ &\leq \left\| \xi_0 - \eta_0 \right\| + \int_{t_0}^{t} \left(\alpha(\tau) \left\| u(\tau) - v(\tau) \right\| + \beta(\tau) \right) \mathrm{d}\tau\,, \end{aligned}$$

which is the desired estimation. □

The last lemmas provide two auxiliary tools which are needed to prove the existence of a solution of the IVP as guaranteed by the following fundamental existence theorem.

Theorem 21 (Existence and uniqueness of a solution). *Let $D \subseteq E$ open, $\mathcal{I} \subset \mathbb{R}$ an interval and $f : \mathcal{I} \times D \to E$ be a function which satisfies the assumptions* (F1) *and* (F2)*. Moreover, assume that f fulfills the following properties:*

1. *local Lipschitz property: For every $\xi_0 \in D$ exits a number $\rho > 0$ and a locally integrable function $\alpha \in \mathrm{L}_{\mathrm{loc}}(\mathcal{I}, \mathbb{R}_+)$ such that the ball $\mathcal{B}_\rho(\xi_0) \subseteq D$ and*

$$\|f(\tau, \xi) - f(\tau, \eta)\| \leq \alpha(\tau) \|\xi - \eta\| \quad \forall \xi, \eta \in \mathcal{B}_\rho(\xi_0), \forall \tau \in \mathcal{I}, \tag{2.11}$$

2. *local boundedness: For every fixed $\xi_0 \in D$ there is a locally integrable function $\beta \in \mathrm{L}_{\mathrm{loc}}(\mathcal{I}, \mathbb{R}_+)$ such that*

$$\|f(\tau, \xi_0)\| \leq \beta(\tau) \quad \text{for allmost all } \tau \in \mathcal{I}\,. \tag{2.12}$$

Then there exists for every pair $(t_0, \xi_0) \in \mathcal{I} \times D$ a nonempty subinterval $\mathcal{I}_u \subseteq \mathcal{I}$ which is relative open to $\mathcal{I}$ and a solution u of the IVP (2.3) *on $\mathcal{I}_u$. The solution u is the maximal solution in the interval $\mathcal{I}_u$. In other words, if there is an other solution $v : \mathcal{I}_v \to D$ of the IVP* (2.3) *with $\mathcal{I}_v \subseteq \mathcal{I}$, then*

$$\mathcal{I}_v \subseteq \mathcal{I}_u \quad \text{and} \quad u = v \ \text{ on } \ \mathcal{I}_v\,.$$

Remark 22. The assumptions of the existence theorem imply a stronger condition 2, i.e. the condition 2 holds uniform on compacts. More precisely, there exists for every compact $K \subseteq D$ a locally integrable function $\gamma \in \mathrm{L}_{\mathrm{loc}}(\mathcal{I}, \mathbb{R}_+)$ such that

$$\|f(\tau, \xi)\| \leq \gamma(\tau) \quad \forall \xi \in K\,, \tau \in \mathcal{I}\, a.e. \tag{2.13}$$

Proof of the Remark 22 . Let $\xi_0 \in K$. By the above theorem, there are a number $\rho > 0$ and functions α and β fulfilling the conditions (2.11) and (2.12). It follows that

$$\begin{aligned} \|f(\tau, \xi)\| &\leq \|f(\tau, \xi_0)\| + \|f(\tau, \xi) - f(\tau, \xi_0)\|\,, \\ &\leq \beta(\tau) + \alpha(\tau)\|\xi - \xi_0\|\,, \\ &\leq \beta(\tau) + \rho\alpha(\tau) =: \gamma_{\xi_0}(\tau) \quad \tau \in \mathcal{I}\, a.e., \forall \xi \in \mathcal{B}_\rho(\xi_0)\,. \end{aligned} \tag{2.14}$$

Since α and β are locally integrable, γ_{ξ_0} is locally integrable, too. Because of the compactness of K there is a finite subcover of open balls $\mathcal{B}_\rho$ centered at $\xi_1, \dots \xi_n$. Define the function γ for every $\tau \in \mathcal{I}$ as

$$\gamma(t) := \max\{\gamma_{\xi_1}(\tau), \dots, \gamma_{\xi_n}(\tau)\}\,,$$

which is also locally integrable. By the estimation (2.14), the desired uniform boundedness property (2.13) follows. □

We will now be able to prove the existence theorem.

Proof of Theorem 21 . We will follow the lines of the proof of Theorem 54 in [55], where the statement in proven for the case $E = \mathbb{R}^n$. The proof is based on the application of the Banach contraction theorem as in the standard theory of ordinary differential equations, see e.g. [27, Theorem 2.12].

There is no loss of generality in assuming that $\mathcal{I} \neq \{t_0\}$.We first prove the existence of a solution of the IVP on the interval $[t_0, t_0+\delta] \cap \mathcal{I}$ for some δ. We claim that t_0 is not the right endpoint of $\mathcal{I}$ because then it is obvious. Take $\xi_0 \in D$ and $\rho > 0$. Let $\alpha, \beta \in \mathrm{L}_{\mathrm{loc}}(\mathcal{I}, \mathbb{R}_+)$ locally integrable functions as in the assumptions (2.11) and (2.12). Define two functions

$$a(t) := \int_{t_0}^{t_0+t} \alpha(\tau)\, \mathrm{d}\tau \quad \text{and} \quad b(t) := \int_{t_0}^{t_0+t} \beta(\tau)\, \mathrm{d}\tau$$

and note that $a(t) \to 0$ and $b(t) \to 0$ as $t \to 0^+$ because of the local integrability of α and β. Moreover, since both functions are non negative and non decreasing, there exists a $\delta > 0$ such that $t_0 + \delta \in \mathcal{I}$ and

$$a(t) \leq a(\delta) = \lambda < 1\ \forall t \in [0, \delta] , \tag{2.15}$$

$$a(t)\rho + b(t) \leq a(\delta)\rho + b(\delta) < \rho\ \forall t \in [0, \delta] . \tag{2.16}$$

On the interval $[t_0, t_0+\delta]$ let $u_0(t) \equiv \xi_0$ be constant and $\bar{\mathcal{B}} := \bar{\mathcal{B}}_\rho(u_0)$ be the closed ball around the point u_0 in the space of continuous functions $C([t_0, t_0+\delta], E)$. We define the operator

$$\begin{aligned} T : \bar{\mathcal{B}} &\rightarrow C([t_0, t_0+\delta], E) , \\ (Tu)(t) &:= \xi_0 + \int_{t_0}^{t} f(\tau, u(\tau))\, \mathrm{d}\tau . \end{aligned}$$

Since u is continuous, its image $K := \{u(t) | t \in [t_0, t_0+\delta]\}$ is compact. By Remark 22 there exists a locally integrable function $\gamma \in \mathrm{L}_{\mathrm{loc}}(\mathcal{I}, \mathbb{R}_+)$ such that $\|f(\tau, \xi)\| \leq \gamma(\tau)$ for all $\tau \in [t_0, t_0+\delta]$ and for every $\xi \in K$. Hence,

$$\int_{t_0}^{t} \|f(\tau, u(\tau))\|\, \mathrm{d}\tau \leq \int_{t_0}^{t} \gamma(\tau)\, \mathrm{d}\tau \quad t \in [t_0, t_0+\delta]\ a.e.,$$

and $f(., u(.))$ is locally integrable. Therefore, T is well defined, and Tu is absolutely continuous.

We show that T is a map from $\bar{\mathcal{B}}$ to $\bar{\mathcal{B}}$. To do this, take $u \in \bar{\mathcal{B}}$ and apply Lemma 20 with $\mathcal{I} = [t_0, t_0+\delta]$, $X_0 = \{\xi_0\}$, $X = \mathcal{B}_\rho(\xi_0)$, $\eta_0 = \xi_0$, $v = u_0$, $g \equiv 0$ and same f, α, β and u as here. According to these choices, we have

$$\tilde{u}(t) = (Tu)(t) \quad \text{and} \quad \tilde{v}(t) = u_0 .$$

By the Lemma, we conclude for $t \in \mathcal{I}$, $t \geq t_0$ that

$$\begin{aligned} \|\tilde{u}(t) - u_0(t)\| &\leq \int_{t_0}^{t} (\alpha(\tau) \|u(\tau) - u_0(\tau)\| + \beta(\tau))\, \mathrm{d}\tau , \\ &\leq \|u - u_0\|_\infty \int_{t_0}^{t} \alpha(\tau)\, \mathrm{d}\tau + \int_{t_0}^{t} \beta(\tau)\, \mathrm{d}\tau , \\ &\leq \rho a(\delta) + b(\delta) \underset{(2.16)}{\leq} \rho . \end{aligned}$$

This clearly implies

$$\|Tu - u_0\|_\infty \leq \rho ,$$

which shows that $Tu \in \bar{\mathcal{B}}$.

Let us prove that T is a contraction. To this end, consider $u, v \in \bar{\mathcal{B}}$. We use again Lemma 20, this time with $\mathcal{I} = [t_0, t_0 + \delta]$, $X_0 = X = \mathcal{B}_\rho(\xi_0)$, $g \equiv f$, $\beta \equiv 0$, $\xi_0 = \eta_0$ and α, f and ξ_0 as here. Therefore,

$$\tilde{u}(t) = (Tu)(t) \quad \text{and} \quad \tilde{v}(t) = (Tv)(t)\,,$$

and it follows that

$$\begin{aligned} \|(Tu)(t) - (Tv)(t)\| &\leq \int_{t_0}^{t} \alpha(\tau) \|u(\tau) - v(\tau)\| \, \mathrm{d}\tau\,, \\ &\leq \|u - v\|_\infty a(\delta)\,. \end{aligned}$$

By (2.15) we conclude that

$$\|Tu - Tv\|_\infty \leq \lambda \|u - v\|_\infty \quad \text{with } \lambda < 1\,,$$

which shows the contraction property. By the Banach contraction mapping theorem 60, T has a unique fixpoint $\bar{u} = T\bar{u}$ which is a solution of the IVP on the interval $[t_0, t_0 + \delta]$.

The same conclusion can be drawn for the existence of a solution on $[t_0 - \delta, t_0]$, if t_0 is not a left endpoint of $\mathcal{I}$. By concatenating both solutions, we obtain a solution on the interval $[t_0 - \delta, t_0 + \delta]$. In the case that t_0 is an endpoint, we have either a solution on $[t_0, t_0 + \delta]$ or on $[t_0 - \delta, t_0]$, and in both cases we have a solution defined in a neighborhood of t_0 in $\mathcal{I}$ for any pair (t_0, ξ_0).

In the next step we show the uniqueness of the solution. For this purpose, we suppose that there are two solutions u and v on the interval $\mathcal{I}_v \subseteq \mathcal{I}$, $t_0 \leq t \in \mathcal{I}_v$. We observe the solutions in some interval $\mathcal{J} = [t_0, t_0 + \delta]$ for some $\delta > 0$. Given a u_0, take $\rho > 0$ such that the local Lipschitz property (2.11) holds for f. There is a δ small enough that all values $u(t)$ and $v(t)$ belong to the ball $\mathcal{B}_\rho(\xi_0)$ for $t \in \mathcal{J}$ because of the continuity of u and v. Choosing $X_0 = X$, $\xi_0 = \eta_0$, $g \equiv f$, $\beta \equiv 0$ and f, ξ_0, u, v and α as here, we can deduce again by Lemma 20 that

$$\|(Tu)(t) - (Tv)(t)\| \leq \int_{t_0}^{t} \alpha(\tau) \|u(\tau) - v(\tau)\| \, \mathrm{d}\tau \;\; \forall t \in \mathcal{J}\,.$$

Since u and v are both solutions, they are both fixpoints of T. Therefore, we obtain

$$\|\mu(t)\| \leq \int_{t_0}^{t} \alpha(\tau) \|\mu(\tau)\| \, \mathrm{d}\tau$$

for all $t \in \mathcal{I}$ and $\mu(t) = u(t) - v(t)$. Since $\beta \equiv 0$, the Corollary 19 of Gronwall's Lemma implies that $\mu(t) \equiv 0$, which establish the local uniqueness of the solutions on the interval $\mathcal{J}\,[t_0, t_0 + \delta]$ for any $t_0 \in \mathcal{I}$ and $\xi_0 \in D$.

Suppose that there are two solutions u and v on $\mathcal{I}_v$. Then we could find a $t \in \mathcal{I}_v$ such that $u(t) \neq v(t)$. We define

$$t_1 := \inf\{t \in \mathcal{I}_v, t > t_0 \,|\, u(t) \neq v(t)\}$$

and note that $u \equiv v$ on $[t_0, t_1)$. Since both u and v are continuous, it also holds that $u(t_1) = v(t_1)$. We consider the initial value problem with t_1 as initial time and initial state $\xi_1 := u(t_1)$. By the local uniqueness statement proved just before, we know that $u \equiv v$ on some interval $[t_1, t_1 + \delta]$ for $\delta > 0$. This is a contradiction to the definition of t_1. Hence, we have shown the uniqueness on the interval $\mathcal{I}_v$.

It remains to prove that there is a maximal solution. Therefore, we define

$$t_{\min} := \inf \{t \in \mathcal{I} \mid \exists \, \text{solution on } [t, t_0]\}$$

and

$$t_{\max} := \sup \{t \in \mathcal{I} \mid \exists \, \text{solution on } [t_0, t]\} \ .$$

By the local existence result, it follows $t_{\min} < t_{\max}$. We consider the open interval $(t_{\min}, t_{\max})$ and the sequences

$$s_n \searrow t_{\min} \quad \text{and} \quad v_n \nearrow t_{\max} \, .$$

By definition of $t_{\min}$ and $t_{\max}$, there is a solution on each interval (s_n, v_n) and these solutions coincide on their common domains because of the uniqueness statement. Hence, there is a solution on the interval $(t_{\min}, t_{\max})$. We construct a new interval I_u. If $t_{\min}$ and $t_{\max}$ are interior points of $\mathcal{I}$, we take

$$\mathcal{I}_u := (t_{\min}, t_{\max}) \ .$$

If $t_{\min}$ is the left endpoint of $\mathcal{I}$, we add it to $\mathcal{I}_u$ provided that a solution exists on an interval including $t_{\min}$. Do the same for $t_{\max}$. We thus obtain a nonempty interval which is relatively open to $\mathcal{I}$. Furthermore, if $t_{\min}$ is an interior point of $\mathcal{I}$, then there cannot be any solution v defined on $[t_{\min}, t_0]$ because the local existence statement applied for the IVP with $u(t_{\min}) = v(t_{\min})$ would imply the existence of a solution on some interval $(t_{\min} - \delta, t_{\min}]$ for some $\delta > 0$. Hence, the concatenation of the solutions on the interval $(t_{\min} - \delta, t_0]$ contradicts the definition of $t_{\min}$. Similar arguments apply to $t_{\max}$. That means that any solution of the IVP must have a domain included in $\mathcal{I}_u$, and by the uniqueness statement, it follows the desired maximal solution property. □

We have seen in the proof that the local Lipschitz property for f is needed to show the uniqueness of the solution. The next Proposition provides a criterion for this property.

Proposition 23. *Let $D \subseteq E$ open and $f : \mathcal{I} \times D \to E$ satisfy the assumption* (F1). *Furthermore, suppose that $f(t,.) \in \mathcal{C}^1(D, E)$ for each $t \in \mathcal{I}$ and that there exists some locally integrable function $\gamma \in \mathrm{L}_{\mathrm{loc}}(\mathcal{I}, \mathbb{R}_+)$ for each compact subset $K \subseteq D$ such that*

$$\left\| \frac{\partial f}{\partial \xi}(t, \xi) \right\| \le \gamma(t) \quad \textit{for all } t \in I, \, \xi \in K \, . \tag{2.17}$$

Then f satisfies the local Lipschitz property (2.11) *in Theorem 21.*

Proof. Take $\xi_0 \in D$, $\rho > 0$ such that $\xi_0 \in \mathcal{B}_\rho(\xi_0) \subseteq D$. We denote with K the closed ball $\overline{\mathcal{B}_\rho(\xi_0)}$. By the mean value theorem, we have for any μ, $\eta \in K$ and arbitrary $t \in \mathcal{I}$

$$f(t, \mu) - f(t, \eta) = \frac{\partial}{\partial \xi} f(t, \xi)(\mu - \eta) \quad \text{for } \xi \in \overline{\mu\eta} \in K \, .$$

Hereby, $\overline{\mu\eta}$ denotes the line connecting μ and η. Taking norms and applying the inequality (2.17), we obtain

$$\begin{aligned} \|f(t,\mu) - f(t,\eta)\| &\leq \left\|\frac{\partial}{\partial\xi} f(t,\xi)\right\| \|\mu - \eta\| , \\ &\leq \gamma(\tau) \|\mu - \eta\| \quad \forall \mu,\eta \in K \quad \forall t \in \mathcal{I} , \end{aligned}$$

which completes the proof. □

Proposition 24. *Under the hypotheses of Theorem 21, if moreover there exists a compact set $K \subseteq D$ such that*

$$u(t) \in K \qquad \textit{for all } t \in \mathcal{I}_u$$

holds for the maximal solution u of the IVP (2.3) as defined in the statement of the Theorem 21, then $\mathcal{I}_u = [t_0, \infty) \cap \mathcal{I}$.

Proof. Suppose the assertion of the proposition is false. Then the solution u lives on the interval $[t_0, t_{\max}) \subseteq \mathcal{I}$ with $t_{\max} < \infty$, but there is no solution on $[t_0, t_{\max}]$. According to Remark 22 take a locally integrable function $\gamma : [t_0, t_{\max}] \to \mathbb{R}_+$ for the compact set K in the statement of the proposition. Hence,

$$\|f(t,\xi)\| \leq \gamma(t) \qquad t \in \mathcal{I}\, a.e., \forall \xi \in K .$$

Let $\xi^* \in K$ and the function $\tilde{u}$ be defined by

$$\tilde{u}(t) := \begin{cases} u(t) , & t \in [t_0, t_{\max}) , \\ \xi^* , & t = t_{\max} . \end{cases}$$

Let us examine

$$u(t_{\max}) := \xi_0 + \int_{t_0}^{t_{\max}} f(\tau, \tilde{u}(\tau))\, \mathrm{d}\tau ,$$

which is well-defined because $f(\tau, u(\tau))$ is bounded by the integrable function γ and hence itself integrable. Moreover, it holds that

$$u(t_{\max}) = \xi_0 + \lim_{t \to t_{\max}} \int_{t_0}^{t} f(\tau, \tilde{u}(\tau))\, \mathrm{d}\tau ,$$

and since K is closed, $u(t_{\max}) \in K \subseteq D$. As $f(\tau, \tilde{u}(\tau)) = f(\tau, u(\tau))$ almost everywhere, we can conclude

$$\begin{aligned} u(t) &= \xi_0 + \int_{t_0}^{t} f(\tau, u(\tau))\, \mathrm{d}\tau , \\ &= \xi_0 + \int_{t_0}^{t} f(\tau, \tilde{u}(\tau))\, \mathrm{d}\tau \end{aligned}$$

for $t \in [t_0, t_{\max}]$. We see that u is absolutely continuous from the second line, and we conclude by (INT) from the first line that u is a solution of the IVP on the interval $[t_0, t_{\max}]$, which is a contradiction to our assumption at the begin of the proof. □

The next theorem states that the solution of the IVP depends continuously on the right-hand side and on initial conditions.

Theorem 25 (Continuous dependency on initial conditions and the right-hand side f). *Let $\mathcal{I} = [t_0, t_1]$ a bounded closed interval in $\mathbb{R}$, $D \subseteq E$ open and $C \geq 0$ a real number. Let $\underline{\alpha} \in \mathrm{L}_{\mathrm{loc}}(\mathcal{I}, \mathbb{R}_+)$ be a locally integrable function. Assume that the two functions $f, h : \mathcal{I} \times D \to E$ satisfy the assumptions of Theorem 21, i.e. the conditions* (F1), (F2) *as well as the local Lipschitz and the integrability property* (2.11) *and* (2.12). *Moreover, let $u : \mathcal{I} \to D$ be a solution of the IVP* (2.3) *such that the C-neighborhood*

$$K = \{\xi | \; \|\xi - u(t)\| \leq C \textit{ for some } t \in [t_0, t_1]\}$$

of its range is included in D. Write

$$H(t) := \int_{t_0}^{t} h(\tau, u(\tau)) \; d\tau \,, \qquad t \in \mathcal{I},$$

and $\underline{H} := \sup_{t \in \mathcal{I}} \|H(t)\|$. Suppose that the following conditions are fulfilled:

1.
$$\max\{\underline{H}, \|u(t_0) - \zeta_0\|\} \leq \frac{C}{2} \, e^{-\int_{t_0}^{t_1} \underline{\alpha}(\tau) \, d\tau} \,. \tag{2.18}$$

2. *For $g := f + h$ it holds that*
$$\|g(\tau, \xi) - g(\tau, \zeta)\| \leq \underline{\alpha}(\tau) \, \|\xi - \zeta\| \qquad \forall \xi, \zeta \in D \,, \; \forall \tau \in I \,. \tag{2.19}$$

Under the above assumptions, there exists a solution $v : \mathcal{I} \to D$ of the disturbed IVP

$$\dot{v}(t) \;=\; g(t, v(t)) \,, \qquad v(t_0) = \zeta_0 \,, \tag{2.20}$$

and the solution is uniformly close to u, i.e.

$$\|u - v\|_\infty \leq (\|u(t_0) - \zeta_0\| + \underline{H}) \; e^{\int_{t_0}^{t_1} \underline{\alpha}(\tau) \, d\tau} \,. \tag{2.21}$$

Proof. Our proof starts with the observation that the existence of the solution v of the perturbed IVP (2.20) is a direct consequence of the Existence and Uniqueness Theorem 21. Since we assumed that both f and h satisfy all conditions of this theorem, their sum $g = f + h$ fulfills them, too. For this reason, there exists a unique solution v of the perturbed IVP on the interval $\mathcal{I}_v$.

Our task is to estimate the distance of this solution v to the solution u of the non-disturbed IVP (2.3). Writing

$$\begin{aligned} u(t) - v(t) &= u(t_0) + \int_{t_0}^{t} f(\tau, u(\tau)) \, \mathrm{d}\tau - \zeta_0 - \int_{t_0}^{t} g(\tau, v(\tau)) \, \mathrm{d}\tau \,, \\ &= u(t_0) - \zeta_0 + \int_{t_0}^{t} (g(\tau, u(\tau)) - g(\tau, v(\tau))) \, \mathrm{d}\tau + \int_{t_0}^{t} h(\tau, u(\tau)) \, \mathrm{d}\tau \end{aligned}$$

for $t \in \mathcal{I}$, we conclude that

$$\begin{aligned} \|u(t) - v(t)\| &\leq \|u(t_0) - \zeta_0\| + \int_{t_0}^{t} \|g(\tau, u(\tau)) - g(\tau, v(\tau))\| \, \mathrm{d}\tau + \|H(t)\| \,, \\ &\leq \|u(t_0) - \zeta_0\| + \int_{t_0}^{t} \underline{\alpha}(\tau) \, \|u(\tau) - v(\tau)\| \, \mathrm{d}\tau + \underline{\mathrm{H}} \qquad \forall t \in \mathcal{I} \,, \end{aligned}$$

where we used the estimation (2.19) in the second line. Define the function $\mu : \mathcal{I} \to \mathbb{R}_+$ by $\mu(t) := \|u(t) - v(t)\|$ and rewrite the above inequality as

$$\mu(t) \leq (\|u(t_0) - \zeta_0\| + \underline{\mathrm{H}}) + \int_{t_0}^{t} \underline{\alpha}(\tau)\mu(\tau)\,\mathrm{d}\tau\,, \qquad t \in \mathcal{I}\,.$$

By the Corollary 19 of Gronwall's Lemma, it follows that

$$\mu(t) \leq (\|u(t_0) - \zeta_0\| + \underline{\mathrm{H}})\, \mathrm{e}^{\int_{t_0}^{t} \underline{\alpha}(\tau)\,\mathrm{d}\tau} \leq (\|u(t_0) - \zeta_0\| + \underline{\mathrm{H}})\, \mathrm{e}^{\int_{t_0}^{t_1} \underline{\alpha}(\tau)\,\mathrm{d}\tau}$$

for all $t \in \mathcal{I}$. By this inequality, we can draw two conclusions. Taking the supremum for $t \in \mathcal{I}$ establishes the desired estimation (2.21). Particularly, by applying the assumption (2.18), we conclude that

$$\mu(t) = \|u(t) - v(t)\| \leq C \quad \text{for all } t \in \mathcal{I}$$

and hence that $v(t) \in K \subseteq E$. Therefore, it follows by Proposition 24 that $\mathcal{I}_v = \mathcal{I}$ as claimed. □

2.3.2 Initial value problems with distributed parameter

In this Section we want to investigate initial value problems where the right-hand side depends on distributed parameters. In particular, we are interested in the dependency of the solution of the IVP on this parameter because this characterizes the behavior of the so-called parameter-to-state-map which we want to invert in Chapter 4. These kind of problems are closely related to the theory of control problems where the parameter plays the role of the control function.

Let us consider the following initial value problem

$$\dot{u}(t) = f(t, u(t), p(t))\,, \qquad u(t_0) = \xi_0. \tag{2.22}$$

We will later assume that the function f is continuous in p and u and differentiable in u. The question arises which regularity in t we have to assume for f, if we wish the parameter to be locally essentially bounded. To answer this question, consider the typical situation where you like to compare the trajectory (u, p) with a trajectory (v, q) to another parameter q. Thus, one is interested in the deviations $h = u - v$, $y = p - q$. The deviation h is solution of the IVP

$$\dot{h}(t) = g(t, h(t), y(t))\,,$$

where g is given by

$$g(t, \xi, \omega) := f(t, \xi + v(t), \omega + q(t)) - f(t, v(t), q(t))\,.$$

Observe that, even if f is independent of t, the function g will depend on t in the way that a continuous function v and an essential bounded function q are substituted into f. This motivates the following assumption for f.

In the remainder of this section, we will require that $D \subseteq E$ is open in the n-dimensional Banach space E, that P is a normed space and that f is a function

$$f : \mathbb{R} \times D \times P \to E$$

which is defined by

$$f(t,\xi,\omega) = \tilde{f}(\pi(t),\xi,\omega) ,$$

where

$$\tilde{f} : S \times D \times P \to E , \quad \pi : \mathbb{R} \to S$$

and S is another normed space. The requirements on $\tilde{f}$ are:

1. For each fixed $s \in S$ and $p \in P$ holds that
$$\tilde{f}(s,.,p) \in C^1(D,E) .$$
2. Both $\tilde{f}$ and $\tilde{f}_\xi$ are continuous on $S \times D \times P$.
3. The function π is measurable and locally essentially bounded, i.e. $\pi \in L^\infty_{\text{loc}}(\mathcal{I},S)$.

Remark 26. If the function f is independent of t, the assumption is fulfilled if it holds for $f : D \times P \to E$ that

1. for fixed $p \in P$
$$f(.,p) \in C^1(D,E) .$$
2. both f and f_ξ are continuous on $D \times P$.

These assumptions are sufficient to ensure the existence of a solution of the parameter-dependent IVP like stated in the next theorem.

Theorem 27 (Existence of the solution for the parameter-dependent IVP). *Under the assumptions of this section, there exists for any real number $t_0 < t_{end}$ and every measurable essentially bounded function $p \in L^\infty(\mathcal{I},P)$ defined on $\mathcal{I} := [t_0,t_{end}]$, a nonempty subinterval $\mathcal{I}_u \subseteq \mathcal{I}$ containing t_0, which is relatively open in $\mathcal{I}$. Moreover, the parameter-dependent IVP* (2.22) *has a solution $u : \mathcal{I}_u \to D$. This solution is unique and maximal, i.e. if there is another solution $v : \mathcal{I}_v \to D$ of the IVP with $\mathcal{I}_v \subseteq \mathcal{I}$, then*

$$\mathcal{I}_v \subset \mathcal{I}_u \qquad \textit{and} \qquad u = v \textit{ on } \mathcal{I}_v .$$

The parameter p is called admissible *for ξ_0, if $\mathcal{I}_v = \mathcal{I}$.*

Proof. The main idea of the proof is to apply the Existence Theorem 21 for the IVP without parameter. Therefore, we define the function $g : \mathcal{I} \times D \to E$ by

$$g(t,\xi) := f(t,\xi,p(t)) = \tilde{f}(\pi(t),\xi,p(t)) .$$

Since $\tilde{f}$ is continuous, g is continuous and fulfills the condition (F2). Moreover, g is the composition of a continuous and a measurable function $\varphi : \mathcal{I} \to P \times S$, $t \mapsto (p(t),\pi(t))$, and hence g itself is measurable and satisfies condition (F1).

In order to apply the Existence theorem, it remains to show that g fulfills the local Lipschitz and integrability condition. We claim that there exist compact sets K_1 and K_2 such that $p(t) \in K_1$ and $\pi(t) \in K_2$ for almost all $t \in \mathcal{I}$ because we assumed that p is bounded and that π is locally bounded. If $K \subseteq D$ is also compact, then by continuity of $\tilde{f}$ and $\tilde{f}_\xi$ on $K_2 \times K \times K_1$ it follows that there are upper bounds M_1 and M_2 such that

$$\|g(t,\xi_0)\| = \left\|\tilde{f}(\pi(t),\xi_0,p(t))\right\| \leq M_1 \quad \text{for allmost all}\, t \in \mathcal{I}$$

and

$$\|g_\xi(t,\xi)\| = \left\|\tilde{f}_\xi(\pi(t),\xi,p(t))\right\| \leq M_2 \quad \text{for allmost all}\, t \in \mathcal{I}\,.$$

By Proposition 23, g fulfills the local Lipschitz property (2.11). This allows us to apply the Existence Theorem 21 of the previous section for g, which completes the proof. □

Definition 28. Let f satisfy the assumptions of this section and define the set

$$\mathcal{D}_\phi := \{(t_1,t_0,\xi,p)\,|t_0 < t_1, \xi \in D, p : [t_0,t_1) \to P \text{is admissible for}\,\xi\}\,.$$

The function $\phi : \mathcal{D}_\phi \to D$ is defined by

$$\phi(t_1,t_0,\xi,p) := u(t_1)\,,$$

where u is the unique solution of the IVP (2.22) with initial condition $u(t_0) = \xi$ on the interval $[t_0,t_1]$.

Our aim is to analyze the function ϕ and the conditions which are necessary for $\phi(t_1,t_0,.,.)$ to be continuous or continuously differentiable. Therefore, we consider for any fixed pair $t_0 < t_1$ in $\mathcal{I}$ the sets

$$\mathcal{D}_{t_0,t_1} := \{(\xi,p)\,|\,(t_1,t_0,\xi,p) \in \mathcal{D}_\phi\} \subseteq D \times L^\infty(\mathcal{I},P)$$

and

$$\mathcal{D}_{t_0,t_1,\xi} := \{p\,|\,(t_1,t_0,\xi,p) \in \mathcal{D}_\phi\} \subseteq L^\infty(\mathcal{I},P)\,.$$

Both sets are domains of continuous functions mapping the parameter and the initial condition to the final state and the entire solution like shown in the next theorem. The statement can be found in [55, first part of Theorem 1] for the case of $E = \mathbb{R}^n$ and $P = \mathbb{R}^m$.

Theorem 29. *Under the assumptions of this section consider the autonomous IVP, i.e. the right-hand side f is independent of t. Moreover, define the map*

$$\alpha : \mathcal{D}_{t_0,t_1} \to D\,, \qquad (\xi,p) \mapsto u(t_1)\,,$$

which connects initial condition and parameter to the final state, as well as the function mapping to the entire solution by

$$\psi : \mathcal{D}_{t_0,t_1} \to C^0(\mathcal{I},D)\,, \qquad (\xi,p) \mapsto u\,,$$

where the solution u is given by $u(t) = \phi\left(t,t_0,\xi,p|_{[t_0,t_1)}\right)$ for all $t \in [t_0,t_1]$. Then the following assertions hold:

1. *$\mathcal{D}_{t_0,t_1}$ is open in $E \times L^\infty(\mathcal{I},P)$, and both α and ψ are continuous.*
2. *Given ξ and p admissible for ξ, write $u := \psi(\xi,p)$. Assume that $\{p_j\}_{j=1}^\infty$ is an equibounded sequence of parameters and*

$$\lim_{j\to\infty} \xi_j = \xi\,.$$

If $p_j \to p$ pointwise almost everywhere, then $u_j = \psi(\xi_j,p_j)$ is defined for every j large enough and

$$\lim_{j\to\infty} u_j = u\,.$$

Proof. The set $\mathcal{D}_{t_0,t_1}$ is open in $E \times L^\infty(\mathcal{I}, P)$ because the set of admissible parameters p is open for fixed t_0 and t_1. The continuity of α and ψ follows from the conclusion 2 of the theorem. Therefore, assume that $p_j \to p$ uniformly in $L^\infty([t_0, t_1], P)$. Then in particular the sequence converges pointwise almost everywhere. By Lemma 63 it follows that $\{p_j\}_{j=1}^\infty$ is equibounded, i.e. there is some compact set $K \subseteq P$ such that $p_j(t) \in K$ for all j and almost all $t \in [t_0, t_1]$. By the statement 2 of the theorem, $u_j = \psi(\xi_j, p_j)$ is well-defined for large j and $u_j \to u$ for $j \to \infty$, and hence ψ and α are continuous.

Let us now prove the statement 2, which can be shown with help of the Theorem 25. Therefore, take any $(\xi, p) \in \mathcal{D}_{t_0,t_1}$ and an equibounded and pointwise convergent sequence $p_j \to p$ in $L^\infty(\mathcal{I}, P)$ and denote $u = \psi(\xi, p)$. Choose a $D_0 \subseteq D$ open such that $\bar{D}_0 \subseteq D$ is compact and $u(t) \in D_0$ for all $t \in [t_0, t_1]$. Let $K \subseteq P$ be a compact subset such that $p_j(t) \in K$ and $p(t) \in K$ for almost all $t \in \mathcal{I} = [t_0, t_1]$. The existence of such K owes to the equiboundedness of $\{p_j\}_{j=1}^\infty$.

Define now the maps $h_j : \mathcal{I} \times D \to E$ by

$$h_j(t, \eta) := f(\eta, p_j(t)) - f(\eta, p(t)) \qquad \forall j = 1, 2, \ldots$$

and denote

$$\tilde{f}(t, \eta) := f(\eta, p(t)) .$$

Recall that we assumed f to be differentiable in η and both f and f_η to be continuous on $D \times P$, i.e. there are upper bounds M_1 and M_2 such that $\|f(\eta, p)\| \leq M_1$ and $\|f_\eta(\eta, p)\| \leq M_2$ on $\bar{D}_0 \times K$. Thus, it follows that $\|h_j(t, \eta)\| \leq 2M_1$ and $\left\|\frac{\partial}{\partial \eta} h_j(t, \eta)\right\| \leq 2M_2$ for almost all $t \in \mathcal{I}$ and all j. Moreover, h_j is continuous in η and measurable in t by Lemma 62. Consequently, h_j satisfies all assumptions of Theorem 21, i.e. the properties (F1), (F2), (2.11), and (2.12). In the same manner we see that also $\tilde{f}$ fulfills these conditions (compare with the function g in the proof of Theorem 27).

Introduce for each j the function $g_j = \tilde{f} + h_j$. Clearly, $g_j(t, \eta) = f(\eta, p_j(t))$, and hence it follows for $\eta_1, \eta_2 \in D_0$, all j and all $t \in \mathcal{I}$ that

$$\begin{aligned} \|g_j(t, \eta_1) - g_j(t, \eta_2)\| &= \|f(\eta_1, p_j(t)) - f(\eta_2, p_j(t))\| \\ &\leq \|f_\eta(\eta_2, p_j(t))\| \, \|\eta_1 - \eta_2\| \leq M_2 \|\eta_1 - \eta_2\| \end{aligned}$$

because of the boundedness of f_η on $\bar{D}_0 \times K$. Like in the Theorem 25, define

$$H_j(t) := \int_{t_0}^{t} h_j(\tau, u(\tau)) \, \mathrm{d}\tau, \qquad (t \in I),$$

and $\underline{H}_j := \sup_{t \in I} \|H_j(t)\|$. Let us suppose that $\underline{H}_j$ converges uniformly to zero, i.e. that

$$\lim_{j \to \infty} \underline{H}_j = 0 . \tag{2.23}$$

We now apply the Theorem 25 about continuous dependency on initial conditions with $D = D_0$, $f = \tilde{f}$, $\underline{\alpha} \equiv a := M_2$ and $\zeta_0 = \xi_j$ and $h = h_j$ for large j. For any $\epsilon > 0$ we choose a C with $0 < C \leq \epsilon$ such that

$$\{\eta | \; \|\eta - u(t)\| \leq C \text{ for some } t \in \mathcal{I}\} \subseteq D_0$$

and a j_0 such that for all $j \geq j_0$

$$\|\xi - \xi_j\| \leq \frac{\epsilon}{2} \mathrm{e}^{-a(t-t_0)} \quad \text{and} \quad \underline{H}_j \leq \frac{\epsilon}{2} \mathrm{e}^{-a(t-t_0)} .$$

By this choice, we can estimate that

$$\max \left\{ \underline{H}_j, \|\xi - \xi_0\| \right\} \leq \frac{\epsilon}{2} \mathrm{e}^{-a(t-t_0)} \leq \frac{C}{2} \mathrm{e}^{-a(t-t_0)} ,$$

and hence all assumption of the Theorem 25 are fulfilled which implies that

$$\|u - u_j\|_{\infty} \leq \left(\underline{H}_j + \|\xi - \xi_0\| \right) \mathrm{e}^{a(t-t_0)} \leq C \leq \epsilon .$$

Since $\epsilon > 0$ is arbitrary, we showed the desired assertion $\|u - u_j\|_{\infty} \leq \epsilon$.

To complete the proof, it remains to show the assumption (2.23). Suppose that $p_j(t) \to p(t)$ a.e. as $j \to \infty$. Because of the continuity of f, it follows that

$$h_j(t, u(t)) \to 0 \text{ a.e.}$$

Recall that $\|h_j(t, u(t))\| \leq 2M_1$ on $\bar{D}_0 \times K$, and thus

$$\int_{t_0}^{t} \|h_j(\tau, u(\tau))\| \, \mathrm{d}\tau \to 0$$

because of the Lebesgue Dominated Convergence Theorem. This convergence establishes the assumption (2.23). □

In order to prepare the differentiability theorem, we suppose that P is an open subset of a finite dimensional Banach space F. Moreover, we assume that $f \in C^{0,1}(J \times (D \times P), E)$ and that the solution $\phi : D_\phi \to D$ as in Definition 28 is twice continuously differentiable. We now are able to differentiate the parameter-dependent IVP

$$\begin{aligned} \dot{u}(t) = \frac{\partial \phi}{\partial t}(t, t_0, \xi_0, p) = f(t, \phi(t, t_0, \xi_0, p), p(t)) , \\ u(t_0) = \xi_0 \end{aligned} \tag{2.24}$$

with respect to t_0, ξ_0 and p. If we interchange these derivatives with $\partial/\partial t$, we arrive at

$$\begin{aligned} \frac{\partial}{\partial t}\left(\frac{\partial}{\partial t_0}\phi(t, t_0, \xi_0, p)\right) &= \frac{\partial}{\partial t_0}\left(\frac{\partial}{\partial t}\phi(t, t_0, \xi_0, p)\right), \\ &= \frac{\partial}{\partial t_0} f\left(t, \phi(t, t_0, \xi_0, p), p(t)\right) , \\ &= \frac{\partial}{\partial \phi} f\left(t, \phi(t, t_0, \xi_0, p), p(t)\right) \frac{\partial \phi}{\partial t_0}\left(t, t_0, \xi_0, p\right) \end{aligned}$$

and

$$\begin{aligned} 0 &= \frac{\partial}{\partial t_0}\xi_0, \\ &= \frac{\partial}{\partial t_0}\phi(t_0, t_0, \xi_0, p) \\ &= \frac{\partial}{\partial t}\phi(t, t_0, \xi_0, p)\bigg|_{t=t_0} + \frac{\partial}{\partial t_0}\phi\left(t, t_0, \xi_0, p\right)\bigg|_{t=t_0}, \\ &= f(t, \phi(t, t_0, \xi_0, p), p(t))|_{t=t_0} + \frac{\partial}{\partial t_0}\phi\left(t, t_0, \xi_0, p\right)\bigg|_{t=t_0}, \\ &= f(t_0, \xi_0, p(t_0)) + D_2\phi\left(t_0, t_0, \xi_0, p\right) . \end{aligned}$$

Thus, we obtain the function

$$y : J(t_0, \xi_0, p) \to E, \qquad t \mapsto D_2\phi(t, t_0, \xi_0, p)$$

satisfying the IVP

$$\begin{aligned} \dot{y} &= D_2 f\left(t, \phi(t, t_0, \xi_0, p), p(t)\right) y\,, \\ y(t_0) &= -f\left(t_0, \xi_0, p(t_0)\right)\,. \end{aligned} \tag{2.25}$$

In the same way, we can derive the functions

$$z : J(t_0, \xi_0, p) \to \mathcal{L}\left(E, E\right), \qquad t \mapsto D_3\phi(t, t_0, \xi_0, p)$$

satisfying the IVP

$$\begin{aligned} \dot{z} &= D_2 f\left(t, \phi(t, t_0, \xi_0, p), p\right) z\,, \\ z(t_0) &= I_E \end{aligned} \tag{2.26}$$

and

$$v : J(t_0, \xi_0, p) \to \mathcal{L}\left(E, F\right), \qquad t \mapsto D_4\phi(t, t_0, \xi_0, p)$$

being a solution of the IVP

$$\begin{aligned} \dot{v} &= D_2 f\left(t, \phi(t, t_0, \xi_0, p), p(t)\right) v + D_3 f(t, \phi(t, t_0, \xi_0, p), p(t))\,, \\ z(t_0) &= 0\,. \end{aligned} \tag{2.27}$$

Therefore, each partial derivative is again a solution of a linearized differential equation. In the next theorem we will see that this holds also for the differentials of the maps α and ψ defined in the Theorem 29.

Theorem 30 (Differentiability of the solution). *Under the hypotheses of the previous Theorem 29, if moreover P is an open subset of the Banach space F, $f \in C^1\left(D \times P, E\right)$ and (u, p) an arbitrary trajectory, then consider the solution $w : \mathcal{I} \to E$ of the variational equation*

$$\begin{aligned} \dot{w}(t) &= D_1 f\left(u(t), p(t)\right) w(t) + D_2 f\left(u(t), p(t)\right) q(t) \\ w(t_0) &= \eta_0 \end{aligned} \tag{2.28}$$

for $\eta_0 \in D$ and $q \in L^\infty\left(\mathcal{I}, P\right)$. Then it holds that $\alpha \in C^1\left(\mathcal{D}_{t_0,t_1}, D\right)$, $\Psi \in C^1\left(\mathcal{D}_{t_0,t_1}, C^0\left(\mathcal{I}, D\right)\right)$, and

$$\nabla\Psi[\xi, p]\left(\eta_0, q\right) = w\,.$$

In particular, $\alpha\left(\xi, \cdot\right) : D_{t_0,T,\xi} \to D$ is regular at p if the linear map $A : L^\infty\left(\mathcal{I}, P\right) \to D$,

$$q \mapsto \int_{t_0}^{t_1} \Phi(T, \tau) B(\tau) q(\tau)\, d\tau\,,$$

is onto. Hereby, $\Phi(t, \tau)$ denotes the evolution operator of the homogeneous equation associated to $D_1 f\left(u(t), p(t)\right)$ and $B(t) = D_2 f\left(u(t), p(t)\right)$.

Proof. Given (ξ, p), take $u = \psi\left(\xi, p\right)$. Let $g : D \times P \to \mathbb{R}$ a smooth function with compact support and $g \equiv 1$ in a neighborhood $\mathcal{U}$ around the set of values$(u(t), p(t))$. If we now multiply f with this function, it is sufficient to prove the theorem for this new f which we denote by f, too. The new f is bounded and globally Lipschitz-continuous:

$$\left\| f\left(\xi, \omega\right) - f\left(\eta, v\right) \right\|_E \leq M\left(\left\| \xi - \eta \right\|_E + \left\| \omega - v \right\|_P \right)$$

for all $\xi, \eta \in D$ and $\omega, \upsilon \in P$. In particular, the solutions of the IVP (2.22) are defined globally. Expanding $f(\xi+\eta, \omega+\upsilon)$ in a Taylor-series of first order, we obtain

$$f(\xi+\eta,\omega+\upsilon) = f(\xi,\omega) + D_1 f(\xi,\omega)\,\eta + D_2 f(\xi,\omega)\upsilon + R_2(\xi,\eta,\omega,\upsilon) \qquad (2.29)$$

with remainder which can be estimated because of the Mean Value Theorem by

$$R_2(\xi,\eta,\omega,\upsilon) = \int_0^1 [D_1 f(\xi+\tau\eta,\omega+\tau\upsilon) - D_1 f(\xi,\omega)]\,\eta\,\mathrm{d}\tau + \int_0^1 [D_2 f(\xi+\tau\eta,\omega+\tau\upsilon) - D_2 f(\xi,\omega)]\,\upsilon\,\mathrm{d}\tau\,.$$

Note that $R_2(\xi,\eta,\omega,\upsilon)$ is continuous and vanishes for $\eta, \upsilon = 0$. Since R_2 has a compact support in ξ and η, R_2 is uniformly continuous in ξ and η. Therefore, the supremum $C(\eta,\upsilon)$ over all $\xi \in D$ and $\omega \in P$ of

$$\int_0^1 \|D_1 f(\xi+\tau\eta,\omega+\tau\upsilon) - D_1 f(\xi,\omega)\| + \|D_2 f(\xi+\tau\eta,\omega+\tau\upsilon) - D_2 f(\xi,\omega)\|\,\mathrm{d}\tau$$

is continuous and vanishes for $\eta, \upsilon = 0$.

Choose a parameter $\mu \in L^\infty(\mathcal{I}, P)$ sufficiently near p and a $\eta \in D$ near ξ and denote the corresponding trajectory by $v = \psi(\eta,\mu)$. Define $q(t) := \mu(t) - p(t)$ and compute the derivative of $\delta(t) := v(t) - u(t)$ using (2.29), i.e.

$$\begin{aligned}\dot\delta(t) &= f(u(t)+\delta(t), p(t)+q(t)) - f(u(t),p(t))\,,\\ &= D_1 f(u(t),p(t))\,\delta(t) + D_2 f(u(t),p(t))\,q(t) + \gamma(t)\,,\\ &= A(t)\delta(t) + B(t)q(t) + \gamma(t)\,,\end{aligned} \qquad (2.30)$$

where we introduced the abbreviations

$$A(t) = D_1 f(u(t),p(t))\,, \quad B(t) = D_2 f(u(t),p(t))$$

and $\gamma(t) = R_2(u(t),\delta(t),p(t),q(t))$. Thus, we obtained a linear differential equation

$$\dot\delta(t) = A(t)\delta(t) + B(t)q(t) + \gamma(t)$$

with initial condition

$$\delta(t_0) = v(t_0) - u(t_0) = \eta - \xi\,.$$

As a bound for γ we can derive

$$\begin{aligned}\|\gamma(t)\| \le& \int_0^1 \|D_1 f(u(t)+\tau\delta(t), p(t)+\tau q(t)) - D_1 f(u(t),p(t))\|\,\|\delta(t)\|_P\,\mathrm{d}\tau\\ &+ \int_0^1 \|D_2 f(u(t)+\tau\delta(t), p(t)+\tau q(t)) - D_2 f(u(t),p(t))\|\,\|q(t)\|_P\,\mathrm{d}\tau\,,\\ &\le (\|\delta(t)\|_P + \|q(t)\|_P)\cdot\\ &\int_0^1 \|D_1 f(u(t)+\tau\delta(t), p(t)+\tau q(t)) - D_1 f(u(t),p(t))\|\\ &\quad + \|D_2 f(u(t)+\tau\delta(t), p(t)+\tau q(t)) - D_2 f(u(t),p(t))\|\,\mathrm{d}\tau\,,\\ &\le (\|\delta(t)\|_P + \|q(t)\|_P)\,C(\delta(t),q(t))\,.\end{aligned} \qquad (2.31)$$

Assume that w is the solution of (2.28) with $\eta_0 = \eta - \xi$, i.e.

$$\dot{w}(t) = A(t)w(t) + B(t)q(t)\,.$$

Thus, w is the homogenous solution of (2.30). By the variation of constants formula (B.6), the inhomogeneous solution is given by

$$\begin{aligned}
\delta(t) &= \Phi(t,t_0)\,(\eta - \xi) + \int_{t_0}^{t} \Phi\,(t,\tau)\,[B(\tau)q(\tau) + \gamma\,(\tau)]\;\mathrm{d}\tau\,, \\
&= \Phi(t,t_0)\eta_0 + \int_{t_0}^{t} \Phi\,(t,\tau)\,B(\tau)q(\tau)\;\mathrm{d}\tau + \int_{t_0}^{t} \Phi\,(t,\tau)\,\gamma(\tau)\;\mathrm{d}\tau\,, \\
&= w(t) + \int_{t_0}^{t} \Phi\,(t,\tau)\,\gamma(\tau)\;\mathrm{d}\tau\,,
\end{aligned}$$

where Φ is the fundamental solution associated to $A\,(\cdot)$. Since Φ is a linear continuous operator and hence bounded, it follows with (2.31) that

$$\begin{aligned}
\|\delta(t) - w(t)\| &\leq \int_{t_0}^{t} \|\Phi\,(t,\tau)\|\;\|\gamma(\tau)\|\;\mathrm{d}\tau\,, \\
&\leq c_1 \int_{t_0}^{t} (\|\delta(\tau)\|_P + \|q(\tau)\|_P)\,C\,(\delta(\tau), q(\tau))\;\mathrm{d}\tau\,, \\
&\leq c_1\,(\|\delta\|_\infty + \|q\|_\infty)\,\sup_{\tau\in[t,t_0]} C\,(\delta(\tau), q(\tau))\,.
\end{aligned}$$

Using again the Bellman-Gronwall argument, we estimate

$$\|\delta\|_\infty \leq c_2\,(\|\eta_0\|_E + \|q\|_\infty)$$

and can conclude

$$\|\delta(t) - w(t)\| \leq c_3\,(\|\eta_0\|_E + \|q\|_\infty)\,\sup_{\tau\in[t,t_0]} C\,(\delta(\tau), q(\tau))\,.$$

For η near ξ and μ near p, the supremum is small which establishes the desired differentiability.

Finally, the only point remaining concerns the continuous differentiability of Ψ. Therefore, take $(\xi, p) \in D_{t,t_0}$ and $\epsilon > 0$. The task is now to show that there is some $\delta > 0$ such that for each $\left(\tilde{\xi}, \tilde{p}\right) \in D_{t,t_0}$ with

$$\|p - \tilde{p}\| + \left\|\xi - \tilde{\xi}\right\| < \delta \tag{2.32}$$

follows necessarily that

$$\left\|\nabla\Psi\,(\xi, p) - \nabla\Psi\left(\tilde{\xi}, \tilde{p}\right)\right\| < \epsilon\,. \tag{2.33}$$

Set $u = \psi\,(\xi, p)$and $\tilde{u} = \psi\left(\tilde{\xi}, \tilde{p}\right)$ and consider

$$\left\|\nabla\Psi\,(\xi, p)\,(\xi_0, q) - \nabla\Psi\left(\tilde{\xi}, \tilde{p}\right)(\xi_0, q)\right\| = \|w(t) - \tilde{w}(t)\|\,,$$

where w and $\tilde{w}$ are the solutions of

$$\dot{w} = f_u(u,p)\,w + f_p(u,p)q \quad \text{and} \quad \dot{\tilde{w}} = f_u(\tilde{u},\tilde{p})\,\tilde{w} + f_p(\tilde{u},\tilde{p})q\,.$$

If Φ is the fundamental solution of $\dot{w} = f_u(u,p)\,w$ and similar $\tilde{\Phi}$ of $\dot{\tilde{w}} = f_u(\tilde{u},\tilde{p})\,\tilde{w}$ and $B(t) = f_p(u,p)$ and $\tilde{B}(t) = f_u(\tilde{u},\tilde{p})$, we know by the variation of solution formula that

$$w(t) - \tilde{w}(t) = \left(\Phi(t,t_0) - \tilde{\Phi}(t,t_0)\right)\xi_0 + \int_{t_0}^{t}\left[\Phi(t,\tau)\,B(\tau) - \tilde{\Phi}(t,\tau)\,\tilde{B}(\tau)\right]q(\tau)\,\mathrm{d}\tau\,.$$

Because of the definition of the operator norm, we have to show for each $\epsilon > 0$ that there is a $\delta > 0$ for which (2.32) holds such that $\|w(t) - \tilde{w}(t)\| \le \epsilon(\|\xi_0\| + \|q\|)$ for every $\xi_0 \in D$ and $q \in L^\infty(\mathcal{I}, P)$. Note that

$$\begin{aligned}
&\left\|\Phi(t,\tau)\,B(\tau) - \tilde{\Phi}(t,\tau)\,\tilde{B}(\tau)\right\| \\
&\le \left\|\left(\Phi(t,\tau)\left(B(\tau) - \tilde{B}(\tau)\right)\right)\right\| + \left\|\left(\tilde{\Phi}(t,\tau) - \Phi(t,\tau)\right)\tilde{B}(\tau)\right\|, \\
&\le \|\Phi(t,\tau)\|\,\left\|\left(B(\tau) - \tilde{B}(\tau)\right)\right\| + c\left\|\left(\tilde{\Phi}(t,\tau) - \Phi(t,\tau)\right)\right\|,
\end{aligned}$$

where c is any constant bounding the values $\|f_u(\xi,\omega)\|$, whenever ξ is in a neighborhood of $\{u(t)|t \in [t_0,t_1]\}$ and ω in a neighborhood of $\{p(t)|t \in [t_0,t_1]\}$. It follows that

$$\begin{aligned}
&\|w(t) - \tilde{w}(t)\| \\
&\le \left\|\Phi(t,t_0) - \tilde{\Phi}(t,t_0)\right\|\,\|\xi_0\| \\
&\quad + \int_{t_0}^{t}\left(\|\Phi(t,\tau)\|\,\left\|\left(B(\tau) - \tilde{B}(\tau)\right)\right\| + c\left\|\left(\tilde{\Phi}(t,\tau) - \Phi(t,\tau)\right)\right\|\right)\mathrm{d}\tau\,\|q\|_\infty\,.
\end{aligned}$$

Moreover, $\left\|\left(B(\tau) - \tilde{B}(\tau)\right)\right\|$ is small, when (ξ,p) is near $\left(\tilde{\xi},\tilde{p}\right)$ because it is obtained as an evaluation of a continuous matrix function along nearby trajectories. And likewise $\left\|\left(\tilde{\Phi}(t,\tau) - \Phi(t,\tau)\right)\right\|$ is small (consider the variational equation). Thus, the supremum of the integrand is small, which completes the proof. □

To summarize the results of this section, the parameter-dependent initial value problem

$$\dot{u}(t) = f(u(t),p(t))\,, \qquad u(t_0) = u_0 \tag{2.34}$$

has an unique solution $u : \mathcal{I}_u \to D$ for every measurable essentially bounded function $p : \mathcal{I} \to P$ if for $f : D \times U \to E$ holds that $f(.,p) \in C^1(D,E)$ for fixed $p \in P$ and both f and $D_1 f$ are continuous on $D \times P$. If $\mathcal{I}_u = [t_0,t_1] = \mathcal{I}$, then p is admissible. Moreover, the parameter-to-state map

$$\begin{aligned}
\varphi :\ \mathcal{D}_{t_0,T,\xi} &\to AC(\mathcal{I}, D) \\
p &\mapsto u\,,
\end{aligned}$$

where $u(t) = \phi\left(t,T,\xi,p|_{[t_0,T)}\right)$ is the unique solution of (2.34), is continuously differentiable. The derivative is given by $\varphi'(p)q = w$, where w is the solution of the variational equation

$$\dot{w}(t) = D_1 f(u(t),p(t))\,w(t) + D_2 f(u(t),p(t))\,q(t), \qquad w(t_0) = 0\,.$$

These results will be used in Section 4.1 for the analysis of the forward problem.

2.4 Ill-posed problems and regularization

In this section an overview about the theory of inverse problems is given. Inverse problems are usually ill-posed problems such that the solution process must be stabilized by using so-called regularization methods. The concepts and definitions presented in the following are needed in Chapter 4 in order to solve the parameter identification problem for the process model describing an ultra precise turning process.

2.4.1 Inverse problems and ill-posedness

The starting point for every simulation is an abstract model of the technical or physical process under investigation. Such models typically depend on certain system and process parameters. In real life situations some of these parameters are not known precisely or are acceptable only in a small range of manufacturing conditions. In addition, the mathematical formulation of the solution process typically adds some artificial parameters such as step sizes or regularization parameters for stabilizing the numerical solvers. Hence, the system must be calibrated by determining optimal values for those parameters.

Mathematically, the underlying processes are often modeled in form of an operator equation

$$F(x) = y \tag{2.35}$$

with a possibly non-linear operator F mapping between two sets X and Y. The direct problem consists of the computation of the resulting output data y of the process for a given set of parameters x and given model F. For example, assume that the operator F is a model for a rotating machine structure which describes the resulting vibration y of certain machine parts for unbalance distributions x. Knowing the unbalance, the operator F would provide a prediction of the vibrations.

Solving the so-called inverse problem means that for given data y and operator F, the parameter x has to be determined such that $F(x) = y$. In the example of the rotating machine, the task is the determination of the unknown unbalances of the rotating machine parts for given measured vibrations at some sensor positions on the machine casing. Note that in general the parameters may be distributed parameters, i.e. time-depended functions.

Solving the inverse problem may be problematic since it is not guaranteed that the inverse map F^{-1} exits or whether it is continuous. In these cases the problem is called ill-posed according to the following definition, first introduced by Hadamard [25].

Definition 31. [Hadamard] Let $F : X \to Y$ be a map between two topological spaces X and Y. The problem (F, X, Y) is said to be *well-posed* if the following three conditions are fulfilled:

1. there exits for every $y \in Y$ a solution $x \in X$ such that $F(x) = y$,

2. the solution $x \in X$ is unique,

3. the inverse map $F^{-1} : Y \to X$ is continuous, i.e. the solution x depends continuously on the data y.

If one of the conditions is violated, the problem is called *ill-posed.*

In real life applications inverse problems are typically ill-posed. The crucial condition is the third one which becomes in particular a problem if only measured and therefore noisy data y^δ with $\left\|y-y^\delta\right\| \leq \delta$ are available. In the case of an ill-posed model F, small noise on the data can lead to bad reconstructions of the parameters. The condition 3 depends in a crucial way on the spaces X and Y because the continuity of F^{-1} can be forced by the choice of X and Y. However, the two spaces are normally fixed by the application.

A typical ill-posed problem is differentiation as illustrated in the following example.

Example 32 (Differentiation as an ill-posed problem). Set

$$\bar{C}^1(0,1) = \{\phi \in C^1(0,1) | \phi(0) = 0\}$$

and let $A : C(0,1) \to \bar{C}^1(0,1)$,

$$Ax(t) := \int_0^t f(s)\ \mathrm{ds} = y(t), \qquad t \in [0,1]$$

be the integral-operator, which is linear and continuous. Since moreover

$$\|Ax\|_{C^1(0,1)} = \sup_{t\in[0,1]} |Ax(t)| + \sup_{t\in[0,1]} \left|(Ax)'(t)\right| \geq \sup_{t\in[0,1]} |x(t)| = \|x\|_{C(0,1)},$$

the inverse operator exits and is continuous (see for example [48], Theorem 8.1.15). The operator is thus well-posed. But if some noisy data $y^\delta = y + \delta$ shall be differentiated, the noise normally is not differentiable, and therefore it does not belong to $\bar{C}^1(0,1)$. Therefore, the range of A has to be enlarged, for example we have to consider $A : L^\infty(0,1) \to L^\infty(0,1)$. Then the problem is ill-posed. To see this, take $\delta(t) = \varepsilon_n \sin(n\,t)$, $\varepsilon_n > 0$. Thus, it holds that $\left\|y - y^\delta\right\|_{L^\infty} = \sup(\varepsilon_n \sin(n\,t)) = \varepsilon_n$, but for $x^\delta = A^{-1}y^\delta = \left(y^\delta\right)'$ we get

$$\left\|x - x^\delta\right\|_{L^\infty} = \left\|y' - \left(y^\delta\right)'\right\|_{L^\infty} = \sup_n (n\,\varepsilon_n \cos(n\,t)) = n\,\varepsilon_n,$$

i.e. the data error is amplified by the factor n. Setting $\varepsilon_n = 1/\sqrt{n}$ yields

$$\left\|y' - \left(y^\delta\right)'\right\|_{L^\infty} \to 0, \qquad \text{but} \qquad \left\|x - x^\delta\right\|_{L^\infty} \to \infty \qquad \text{for } n \to \infty.$$

The operator A is consequently not continuously invertible and thus ill-posed.

The problem of existence and uniqueness of a solution of (2.35) can be solved by introducing a so called generalized solution, which will be presented briefly in the following. For a detailed description and proofs of the following assertions we refer to [48]. Assume that X and Y are Hilbert spaces and consider a linear and continuous model operator A, i.e.

$$A \in L(X,Y) := \left\{ B : X \to Y \,|\, B \text{ is linear and } \|B\| = \sup_{\|x\|_X=1} \|Bx\|_Y < \infty \right\}.$$

In this setting the problem is ill-posed if the range $\mathcal{R}(A) := \{Ax \,|\, x \in X\}$ is not closed in Y. Instead of solving the problem $Ax = y$, a minimizer of the discrepancy

$$\|Ax - y\|_Y$$

is computed which is equivalent to solve the normal equation

$$A^* Ax = A^* y. \tag{2.36}$$

Here A^* denotes the adjoined of A which is characterized by $\langle Ax|y\rangle_Y = \langle x|A^*y\rangle_X$ for all $x \in X$ and $y \in Y$. If the data $y \in \mathcal{R}(A) \oplus \mathcal{R}(A)^\perp$, there exists a solution of (2.36). To guarantee uniqueness of the solution, among all solutions that one of minimal norm is taken and denoted with $x^\dagger$. This leads to the generalized or pseudo inverse which is defined as the map

$$A^\dagger : \mathcal{D}\left(A^\dagger\right) = \mathcal{R}(A) \oplus \mathcal{R}(A)^\perp \to X, \qquad y \mapsto x^\dagger.$$

This solution concept guarantees the unique existence of a solution if $y \in \mathcal{D}\left(A^\dagger\right)$. However, the pseudo-inverse may still be discontinuous, which is the case if $\mathcal{R}(A)$ is not closed. Moreover, in real applications only measured data $y^\delta$are available which normally do not lie in the domain $\mathcal{D}\left(A^\dagger\right)$ of the pseudo inverse. Thus, the discontinuity and the limited domain of the pseudo-inverse have to be handled. In the next section one of the most common strategies to stabilize the solution process of inverse problems is presented.

2.4.2 Regularization methods

In this subsection a brief introduction into regularization methods are given which are used for the stabilization of the inversion process of ill-posed problems like introduced in the last subsection. Good general references for a detailed overview about regularization theory are [19, 40, 48].

The next definition explains what is meant with a regularization method.

Definition 33. Let $A \in L(X,Y)$ and $\{R_t\}_{t>0}$ a family of continuous (not necessarily linear) maps from Y to X with $R_t 0 = 0$ for all t. If there exits a map $\gamma :]0,\infty[\times Y \to]0,\infty[$ called *parameter choice rule* such that for every $y \in Y$

$$\sup\left\{\left\|A^\dagger y - R_{\gamma(\delta,y^\delta)} y^\delta\right\|_X \,|y^\delta \in Y, \left\|y - y^\delta\right\|_Y \leq \delta\right\} \to 0 \quad \text{as } \delta \to 0, \tag{2.37}$$

the pair $\left(\{R_t\}_{t>0}, \gamma\right)$ is said to be a *regularization method* for $A^\dagger$. It is called *linear regularization* if R_t are linear. The value $\gamma\left(\delta, y^\delta\right)$ is called *regularization parameter* and has to be chosen in the manner that

$$\lim_{\delta \to 0} \sup\left\{\gamma\left(\delta, y^\delta\right) \,|\, y^\delta \in Y, \left\|y - y^\delta\right\|_Y \leq \delta\right\} = 0. \tag{2.38}$$

The above definition provides a quite abstract way how to define reasonable approximations to the generalized solution $x^\dagger$. The reconstruction error $\left\|A^\dagger y - R_t y^\delta\right\|_X$ comprises in the case of a linear regularization two terms, the approximation error and the data error, i.e.

$$\left\|A^\dagger y - R_t y^\delta\right\|_X \leq \left\|A^\dagger y - R_t y\right\|_X + \left\|R_t\left(y - y^\delta\right)\right\|_X. \tag{2.39}$$

The approximation error tends towards zero for $t \to 0$, whereas the data error will explode if the noise $y - y^\delta$ does not belong to $\mathcal{D}(A^\dagger)$ which generally is the case. In contrast, the data error will decrease for $t \to \infty$, but the approximation error increases. Therefore, in order to minimize the reconstruction error, the optimal regularization parameter t_{opt} has to been chosen by balancing the two error terms. This choice should be selected by the parameter choice rule, i.e. $\gamma(\delta, y^\delta) \approx t_{\text{opt}}$.

In the following some examples for classical regularization methods are presented. The first class of regularization methods are filter function methods which can be used for linear compact operators

$$A \in K(X,Y) := \left\{ A \in L(X,Y) \,|\, \overline{AU} \text{ compact in } Y \text{ for every bounded } U \subseteq X \right\}.$$

Inverses of compact operators $A \in K(X,Y)$ with non finite dimensional X are discontinuous and hence are always ill-posed.

Definition 34. Let X and Y are Hilbert spaces and $A \in K(X,Y)$. Then A^*A is self-adjoined in $K(X)$ with eigenvalues λ_n, $|\lambda_1| \geq |\lambda_2| \geq \cdots > 0$ and corresponding eigenvectors v_n. The system $\{(v_n, u_n, \sigma_n) \,|\, n \in \mathbb{N}\} \subset]0,\infty[\times X \times Y$ with

$$\sigma_n := \sqrt{\lambda_n}\,, \qquad u_n := \sigma_n^{-1} A v_n\,, \qquad n \in \mathbb{N},$$

is called *singular system* of A. The series

$$Ax = \sum_{n=1}^{\infty} \sigma_n \langle x \,| v_j \rangle_X \, u_j$$

is said to be the *singular decomposition* of A.

The singular system of A can be used to represent the pseudo-inverse in the following way

$$A^\dagger y = \sum_{n=1}^{\infty} \sigma_n^{-1} \langle y \,| u_n \rangle_X \, v_n \qquad \text{for } y \in \mathcal{D}(A^\dagger)\,.$$

For a proof we refer to [48]. This representation of the pseudo-inverse shows the problematic situation if noisy data y^δ is used for the inversion. The noisy part in direction of u_n is amplified by σ_n^{-1}, and, since $\sigma_n \to 0$, the solution $A^\dagger y$ is useless whether it exits. That's why the stabilization of the inversion process is necessary. The most simplest way to stabilize it, is to stop the expansion of $A^\dagger y$ at some finite index n. This method is called truncated singular value decomposition (TSVD) and is indeed a regularization method, see [48, Theorem 3.3.7 and Example 3.3.9].

Remark 35. The TSVD is part of the general regularization scheme of filter functions F_t which manipulates the singular values. The regularization operators have the form

$$R_t y := F_t(A^*A)\, A^* y = \sum_{n=1}^{\infty} F_t(\sigma_n^2)\, \sigma_n \langle y \,| u_n \rangle_Y \, v_n\,.$$

For the TSVD the filter functions are defined as

$$F_t(\lambda) = \begin{cases} \lambda^{-1}, & \lambda \geq t \\ 0, & \lambda < t \end{cases}$$

which form regularization operators

$$R_t y = \sum_{\sigma_n \geq \sqrt{t}} \sigma_n^{-1} \langle y \,| u_n \rangle_X \, v_n\,.$$

However, in many practical applications the singular value decomposition of an operator is not known or it is only with high time effort determinable. Therefore other strategies for regularization have to be applied. One alternative strategy is based on minimization problems. One advantage of this class of regularization methods is that they can be used for linear as well as nonlinear operator equations.

In the last subsection the minimization of the discrepancy

$$\min_{x \in X} \left\| F(x) - y^\delta \right\|_Y$$

has been introduced as an alternative way to solve the operator equation (2.35) directly. To emphasize the possible problems of this method, note that for ill-posed problems this minimization problem would lead to highly oscillating solutions with infinite energy.

There are several possibilities to avoid this. One is the so-called Landweber regularization based on an iterative algorithm to minimize the squared discrepancy $\Psi(x) = \left\| F(x) - y^\delta \right\|_Y^2$ of the form

$$\begin{aligned} x^{n+1} &= x^n - \mu \nabla \Psi\left(x^n\right), \\ x^0 & \quad \text{arbitrary} \end{aligned}$$

with step size $\mu > 0$. It is also called gradient method because the descent direction is given by the negative gradient of the functional Ψ. For a linear operator $F \equiv A$, it can be shown that the resulting algorithm

$$x^{n+1} = x^n - \mu A^* \left(A x^n - y^\delta\right) \tag{2.40}$$

forms for $0 < \mu < 2/\left\|A\right\|^2$, $x^0 = 0$ and regularization parameter $\alpha = 1/n$ a regularization method, see [48, Theorem 5.1.6]. The parameter choice rule consists here in the determination of the right stopping index n for the iteration (2.40).

Another way to avoid oscillating solutions is the stabilization of the inversion by determining x as the minimizing solution of

$$\min_{x \in X} \left\| F(x) - y^\delta \right\|_2^2 + \alpha\, \Omega(x),$$

where $\Omega(x)$ is called the regularization or penalty term, which controls the energy of x and at the same time models some a priori information on the optimal parameter setting. The regularization parameter $\alpha > 0$ balances the discrepancy and the penalty term, i.e. the choice of α influences the approximation of the data and the penalty term which prevents the solution for exploding indefinitely. The classical setting

$$\Omega(x) = \left\|x\right\|_2^2 = \int x^2(\tau)\, \mathrm{d}\tau \tag{2.41}$$

leads to parameters of minimal energy. The corresponding functional

$$\mathrm{T}_\alpha(x) = \left\| F(x) - y^\delta \right\|_2^2 + \alpha \left\|x\right\|_2^2 \tag{2.42}$$

is the classical Tikhonov-functional which was first investigated by Tikhonov [57, 58]. This type of least squares approach has been investigated since the early 1960's. The Tikhonov-functional is strictly convex, and therefore it has a unique minimizer which is denoted by x_α^δ. If the parameter α is chosen according to an appropriate parameter choice rule, the Tikhonov-method forms a regularization method like stated in the following theorem. The regularization operator is hereby defined as

$$R_\alpha : y^\delta \mapsto x_\alpha^\delta = \operatorname*{argmin}_{x \in X} \mathrm{T}_\alpha(x) .$$

Theorem 36 (Regularization property of the Tikhonov-regularization). *Let* $y \in \mathcal{R}(A)$ *and* $\left\|y - y^\delta\right\| \leq \delta$. *If* α *is chosen such that*

$$\lim_{\delta \to 0} \alpha = 0 \qquad \textit{and} \qquad \lim_{\delta \to 0} \delta^2 / \alpha = 0 , \tag{2.43}$$

then

$$\lim_{\delta \to 0} \left\|x_\alpha^\delta - x^\dagger\right\|_X = 0 .$$

Proof. For a proof, see [48, Theorem 7.4.3]. □

Remark 37. For a linear operator A the Tikhonov-regularization can be expressed as a filter function method with filter function

$$F_\alpha(\lambda) = \frac{1}{\lambda + \alpha} , \qquad \alpha > 0 .$$

Instead of truncate small singular value like for the TSVD-method, the singular values are shifted away from zero by α. Therefore, $R_\alpha y := F_\alpha(A^* A) A^* y$ is the unique solution of the regularized normal equation

$$(A^* A + \alpha I) R_\alpha y = A^* y .$$

The crucial point is still the choice of the optimal regularization parameter. If no information on the minimizer is available, the simplest a priori parameter choice rule satisfying the condition (2.43) in Theorem 36 is to set $\gamma(\delta) = \alpha \sim \delta$. Another popular method to determine α is the discrepancy principle which is an a posteriori rule. The underlying basic idea is to chose $\alpha = \gamma\left(\delta, y^\delta\right)$ such that

$$\left\|A x_\alpha^\delta - y^\delta\right\|_Y \approx \delta ,$$

i.e. such that the discrepancy has the same magnitude as the data error. Searching smaller discrepancies would be useless because one could not expect a higher accuracy than the data error. The advantage of this method is that it provides a concrete scheme for determining the best parameter:

Morozov's Discrepancy principle : Let $\{\alpha_n\}_n$ a monotonically decreasing sequence and $\tau > 1$. Determine n^* such that

$$\left\|A x_{\alpha_{n^*}}^\delta - y^\delta\right\|_Y \leq \tau\delta < \left\|A x_{\alpha_n}^\delta - y^\delta\right\|_Y \qquad n = 1, 2, \ldots, n^* - 1 . \tag{2.44}$$

Set $\gamma\left(\delta, y^\delta\right) = \alpha_{n^*}$.

If a sequence $\alpha_n = q^n \alpha_0$ with $0 < q < 1$ is taken, the iteration can be stopped if the condition (2.44) is fulfilled. No further computation is necessary to check the discrepancy principle because in most of the minimizing schemes the discrepancy is computed as a by-product. There are several generalized discrepancy principles, see [48, Section 3.5].

Heuristic or data driven methods to choose α do not use the noise level δ but only the noisy data y^δ. If the operator equation $Ax = y$ is ill-posed, such heuristic parameter choice rules do not provide a regularization method, see Bakushinskii's veto [4], but can deliver good results in practical applications. Examples are the L-curve criterion or the heuristic method by Hanke and Raus, see [48, Chapter 3.6]. A comparison of different parameter choice rules, a posteriori as well as data driven rules, is provided in [5] and [44].

Tikhonov-regularization can be applied for differentiating noisy data as illustrated in the next example.

Example 38 (Differentiation)**.** Consider again the integral-operator of Example 32. In order to compute the derivative x numerically, the operator is discretized on $[0,1]$. Therefore, a mesh $t_i = (i - 1/2)h$, $i = 1, \dots N$ with mesh-size $h = 1/N$ is defined and $y_i = Ax(t_i)$ is computed by approximating the integral by the rectangle rule, i.e.

$$y_i = Ax(t_i) = \sum_{k=1}^{i-1} \int_{(k-1)h}^{kh} x(s)\ \mathrm{d}s + \int_{(i-1)h}^{t_i} x(s)\ \mathrm{d}s \approx \sum_{k=1}^{i-1} hx(t_k) + \frac{h}{2} x(t_i)\,.$$

Therefore, $\mathbf{y} = (y_i)_{i=1}^N$ can be computed by a matrix-vector multiplication $\mathbf{y} = \mathbf{A}\mathbf{x}$ with $\mathbf{x} = (x(t_i))_{i=1}^N$ and

$$\mathbf{A} = h \begin{pmatrix} 1/2 & 0 & \cdots & 0 \\ 1 & 1/2 & & \vdots \\ \vdots & \ddots & \ddots & 0 \\ 1 & \cdots & 1 & 1/2 \end{pmatrix}.$$

In order to generate noisy data y^δ, consider the function

$$y(t) = -\frac{3}{\pi}\left(\cos\left(\pi t\right) - 1\right)$$

and its derivative $x(t) = 3\sin(\pi t)$. Take $\mathbf{y} = (y(t_i))_{i=1}^n$ and add random Gaussian noise δ in order to generate noisy data $\mathbf{y}^\delta = \mathbf{y} + \delta$, which is plotted in Figure 2.2b.

Although there is no difference of y and y^δ observable with the naked eye, the reconstruction of the derivative obtained by using forward difference quotient is strongly noised. Clearly, regularization must be applied. First, the TSVD-method is used for the stabilization of the differentiation process. The choice of the regularization parameter α influences the quality of the reconstruction, see Figure 2.3.

Next, Tikhonov-regularization is applied in order to differentiate the noisy data. The regularization parameter α is determined automatically by the Morozov principle. The reconstructed derivative x_α^δ as well as the corresponding data $A\left(x_\alpha^\delta\right)$ are shown in Figure 2.4.

Up to now, only the classical choice of the L_2-norm for the penalty term has been considered which leads to the classical Tikhonov-approach. The corresponding minimizing solutions have finite norms and are comparatively smooth because small singular values are damped what weakens the influence of high oscillating components, see Remark 37. In many situations, this smoothing effect is not desirable.

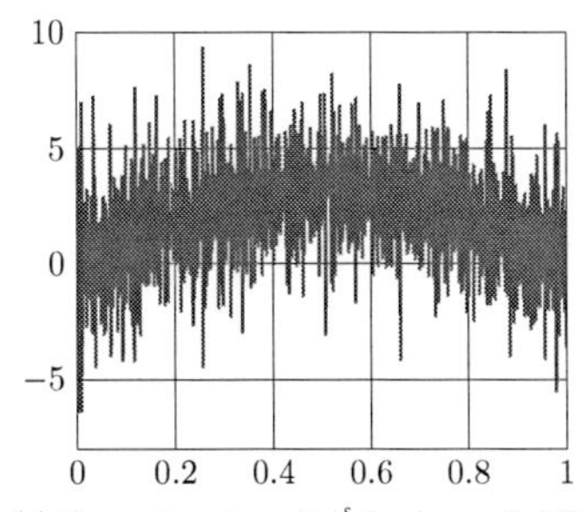

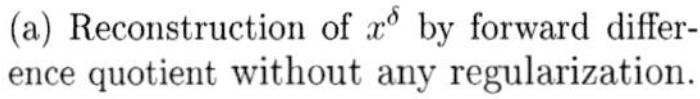

(a) Reconstruction of x^δ by forward difference quotient without any regularization.

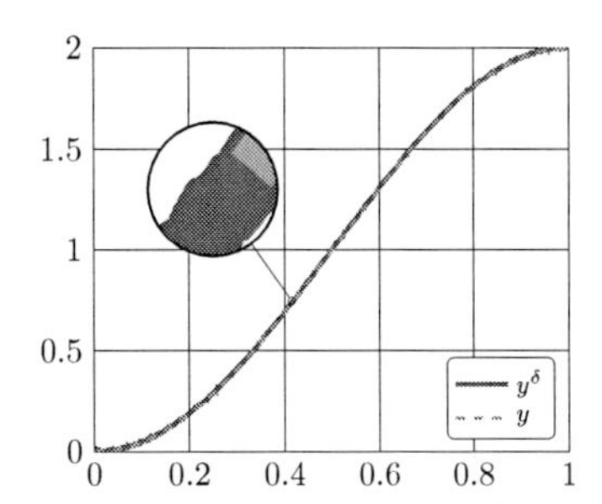

(b) Real data y and noisy data y^δ for $\delta = 0.05$.

Figure 2.2: Naive reconstruction of the derivative x^δ computed with forward difference quotients for given data y^δ. The noise in the data is amplified by the differentiation and leads to a bad reconstruction of the derivative.

For example, thinking of image processing, sharp edges might be preserved when deblurring noisy images. This can be done by Tikhonov-regularization by using as penalty term a TV-norm instead of the squared Hilbert space norm.

Another important class of regularization methods are so-called sparsity regularization. Hereby, the principle interest is not the restriction of the energy of the parameter, but the aim is to determine parameters that are well-localized. For an illustration of this concept, consider a cutting process like for example turning or milling. An optimal time varying feed rate of the cutting tool might be searched in order to obtain a predefined surface y. In addition, the model F allows us to simulate the resulting surface y for given feed rate f. Based on experience, a feed rate f_0 is suggested which however leads to a suboptimal result. The searched parameter in this case is the change of the feed rate, e.g. $f_{opt} = f + x$ and we want to determine x such that $F(f + x) \sim y$. Enforcing the side condition, that one prefers to modify the feed rate f over a minimal time span, leads to a modified Tikhonov-functional. Solutions of this type are obtained with penalty terms of the form

$$\Omega(x) = \int |x(\tau)| \; \mathrm{d}\tau \; . \tag{2.45}$$

This minor change in the formulation of the minimization procedure has a tremendous impact on the structure of the minimizer as well as on the complexity of the numerical schemes needed for approximating the minimizer: the classical L_2-penalty term allows to use Hilbert space techniques leading to least square approximations, which are well developed since decades. In contrast, sparsity promoting L_1-penalty terms require optimization techniques, which were for the first time analyzed in 2004 by [17]. We will discuss these regularization methods in Subsection 2.4.4.

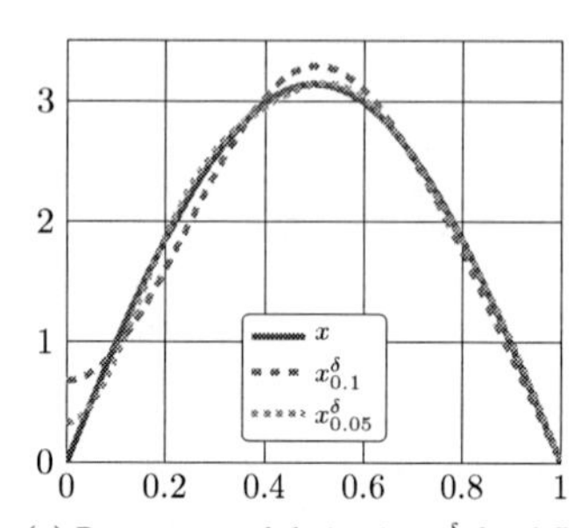

(a) Reconstructed derivative x^δ_α for different regularization parameters α compared to the real solution x.

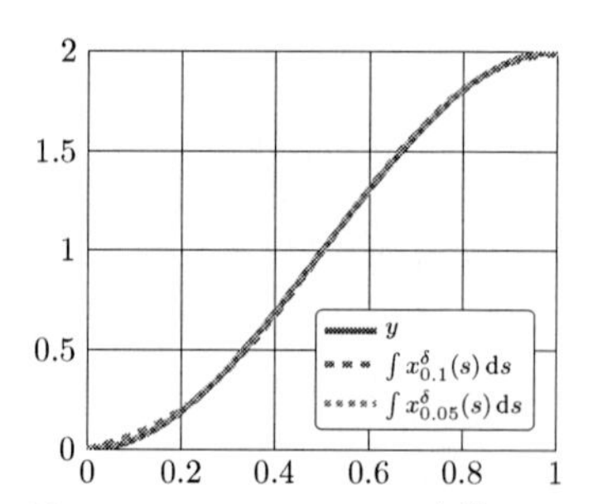

(b) Reconstructed data from $A\left(x^\delta_\alpha\right)$ in comparison to original data y.

Figure 2.3: TSVD-method: Reconstructed derivative x^δ_α and real solution x (left) and comparison of the reconstructed data $A\left(x^\delta_\alpha\right)$ with original data y (right).

2.4.3 Nonlinear inverse problems

In the previous section we have seen that in the case of linear operator equations the problem is ill-posed if the range of the operator is not closed. The question arises how nonlinear problems can be characterized. Therefore, we consider the nonlinear operator $F : \mathcal{D}(F) \subseteq X \to Y$ mapping between two Banach spaces and the corresponding operator equation

$$F(x) = y\,. \tag{2.46}$$

We assume that there is at least one $x^+ \in \mathcal{D}(F)$ such that $F(x^+) = y$.

Definition 39. The nonlinear operator equation (2.46) is said to be *locally ill-posed* in $x^+ \in \mathcal{D}(F)$ if for every $r > 0$ there exists a sequence $\{x_n^r\}_n \subset \mathcal{B}_r(x^+) \cap \mathcal{D}(F)$ such that

$$\lim_{n\to\infty} \left\| F(x_n^r) - F\left(x^+\right) \right\|_Y = 0 \qquad \text{but}\, x_n^r \nrightarrow x^+ \text{ for } n \to \infty\,. \tag{2.47}$$

Otherwise the problem is called *locally well-posed*, i.e. there is a $r > 0$ such that for all sequences $\{x_n^r\}_n \subset \mathcal{B}_r(x^+) \cap \mathcal{D}(F)$ holds

$$\lim_{n\to\infty} \left\| F(x_n^r) - F\left(x^+\right) \right\|_Y = 0 \quad \Rightarrow \quad \lim_{n\to\infty} \left\| x_n^r - x^+ \right\|_X = 0\,.$$

According to the definition, the problem is locally ill-posed if it is not possible to reconstruct a unique solution x^+ locally or like in the case of linear operators if the solution does not depend continuously on noisy data. A linear equation $Ax = y$ is either everywhere locally ill-posed or everywhere locally well-posed. The linear equation is locally ill-posed if A is not injective or the range $\mathcal{R}(A)$ is not closed. In order to find a similar characterization for nonlinear operator F, we need the following two definitions.

Definition 40. The operator $F : \mathcal{D}(F) \subseteq X \to Y$ between normed spaces is called *completely continuous* if it is compact and continuous.

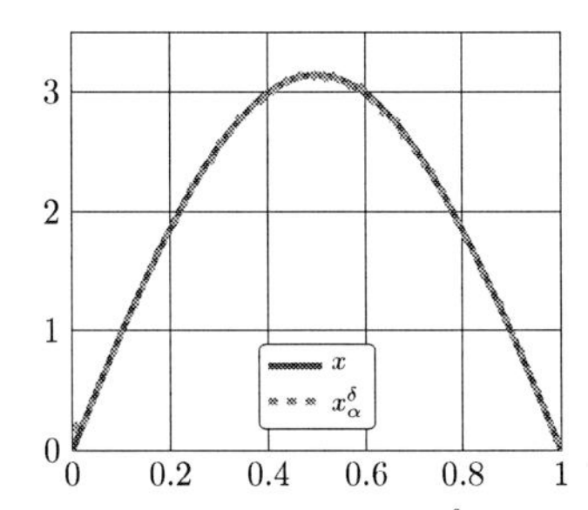

(a) Reconstructed derivative x_α^δ. The regularization parameter α is determined by Morozov's discrepancy principle (2.44).

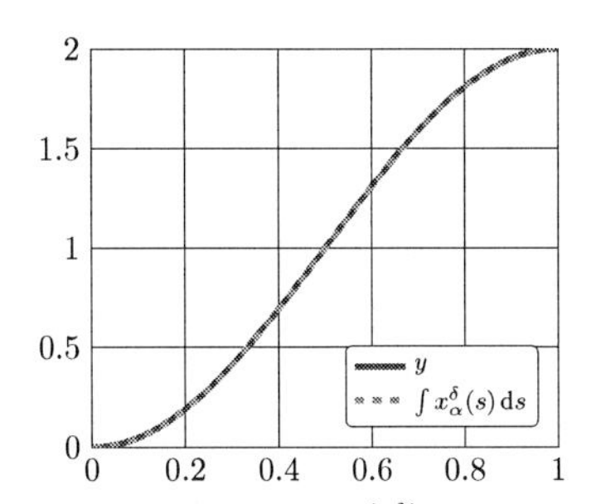

(b) Reconstructed data $A\left(x_\alpha^\delta\right)$ and real data y.

Figure 2.4: Tikhonov-regularization: Reconstructed derivative x_α^δ and real solution x (left) and corresponding data $A\left(x_\alpha^\delta\right)$ and y (right).

Definition 41. The operator $F : \mathcal{D}(F) \subseteq X \to Y$ between Hilbert spaces is called *weakly (sequentially) closed* if for all sequences $\{x_n\}_n \subset \mathcal{D}(F)$ holds that the weak convergence of $x_n \rightharpoonup x$ in X and $F(x_n) \rightharpoonup y$ in Y imply that

$$x \in \mathcal{D}(F) \quad \text{and } F(x) = y\,.$$

Operators which are weakly sequentially closed and completely continuous have the nice property that they transform weakly convergent sequences $x_n \rightharpoonup x$ into strong convergent sequences $F(x_n) \to F(x)$. If moreover X is separable and finite dimensional, the nonlinear operator equation (2.46) is locally ill-posed in $x^+ \in \mathcal{D}(F)$, see [48, Theorem 7.3.4].

As mentioned in the previous subsection, regularization methods based on minimization problems can also be applied for the regularization of nonlinear operator equations. Since the parameter identification problem treated in Chapter 4 can be reformulated such that two linear equations have to be inverted, the regularization theory for nonlinear problems is not discussed in the following. We refer the reader for regularization methods in Hilbert spaces to standard textbooks like [19, 48]. The theory can also formulated in Banach-spaces, see for example [53], where in addition regularization methods with sparsity constraints are discussed. An introduction for sparsity promoting methods for linear operators is given in the next subsection.

2.4.4 Regularization with sparsity constraints

In this section we will explain how to solve inverse problems with the side condition of a sparse solution. Such sparse solutions appear in many inverse problems in signal or image processing or parameter identification problems. In this context, sparsity means that the solution $x^\dagger$of the operator equation

$$F(x) = y$$

with possible nonlinear operator $F : H_1 \to H_2$ mapping between Hilbert spaces can be expressed as the sum of basis or frame elements of $\mathcal{B} = \{\varphi_i\}_{i\in\mathbb{N}}$ in H_1:

$$x^\dagger = \sum_{i\in\mathbb{N}} \langle x^\dagger | \varphi_i \rangle \varphi_i = \sum_{i\in\mathbb{N}} x_i \varphi_i . \tag{2.48}$$

For a sparse solution with respect to the frame or basis $\mathcal{B}$, only a finite number of coefficients x_i are non-zero. For finite dimensional problems a sparse solution should have as few non-zero coefficients as possible.

The knowledge of the sparse structure of the solution can be considered in the Tikhonov-regularization by replacing the classical quadratic penalty term (2.41) with the sparsity enforcing penalty term of the form

$$\Omega_p(x) = \|x\|_{\mathbf{w},p}^{\,p} = \sum_{i\in\mathbb{N}} w_i \, |\langle x | \varphi_i \rangle|^p \tag{2.49}$$

with $1 \leq p \leq 2$ and a sequence $\{w_i\}_{i\in\mathbb{N}}$ of positive weights $w_i \geq w > 0$. This penalty term has first been proposed in 2004 by Daubechies, Defrise and De Mol [17].

A heuristic explanation why solutions of the corresponding Tikhonov-type functional

$$T^p_{\alpha,\mathbf{w}}\left(x, y^\delta\right) = \|F(x) - y\|_2^{\,2} + \alpha \, \|x\|_{\mathbf{w},p}^{\,p} \tag{2.50}$$

are indeed sparse can be given by the following observation. The penalty term (2.49) consists of a sum of p-squared expansion coefficients $|x_i|^p$ regarding to $\mathcal{B}$. Figure 2.5 displays these terms for different p. Note that big coefficients $|x_i| > 1$ are more penalized by $p = 2$ than for smaller $p < 2$, whereas small coefficients $|x_i| < 1$ are less damped for decreasing $p < 2$ than for $p = 2$. With other words, solutions with few big coefficients are less penalized for smaller q as those with many small coefficients in their expansion. Therefore, it can be expected that the Tikhonov-functional (2.50) produces sparse solutions if p is chosen small enough. In the case of $p = 0$ this observation is obvious because for non sparse solutions the Tikhonov-functional would always have the value ∞. But the functional is not convex for $p < 1$, and the analysis and minimization become more complex. Therefore, we will only consider the case $1 \leq p \leq 2$.

The next theorem, also proven first in the pioneering paper [17], shows that the regularization operator $R_\alpha : y^\delta \mapsto x^\delta_\alpha = \operatorname{argmin}_{x\in H_1} T^p_{\alpha,\mathbf{w}}\left(x, y^\delta\right)$ forms indeed a regularization method if F is a linear operator.

Theorem 42 (Regularization property, [17], Theorem 4.1). *Let $F : H_1 \to H_2$ be a linear bounded operator with $\|F\| < 1$, $1 \leq p \leq 2$ and $\{w_i\}_{i\in\mathbb{N}}$ a sequence of weights $w_i \geq w > 0$ bounded uniformly from below. Assume that either $p > 1$ or that the nullspace $\mathcal{N}(F) = \{0\}$. For any $y^\delta \in H_2$ satisfying $\left\|y - y^\delta\right\| \leq \delta$ and any $\alpha > 0$ denote with x^δ_α the minimizer of the Tikhonov-functional $T^p_{\alpha,\mathbf{w}}\left(x, y^\delta\right)$ in* (2.50). *If the regularization parameter $\alpha = \alpha(\delta)$ is chosen such that*

$$\lim_{\delta\to 0} \alpha(\delta) = 0 \quad \text{and} \quad \lim_{\delta\to 0} \frac{\delta^2}{\alpha(\delta)} = 0 ,$$

then it holds that

$$\lim_{\delta\to 0} \left\|x^\delta_\alpha - x^\dagger\right\|_{H_1} = 0,$$

where $x^\dagger$ is defined as the unique minimum $\|\cdot\|_{\mathbf{w},p}$-norm solution of $Fx = y$.

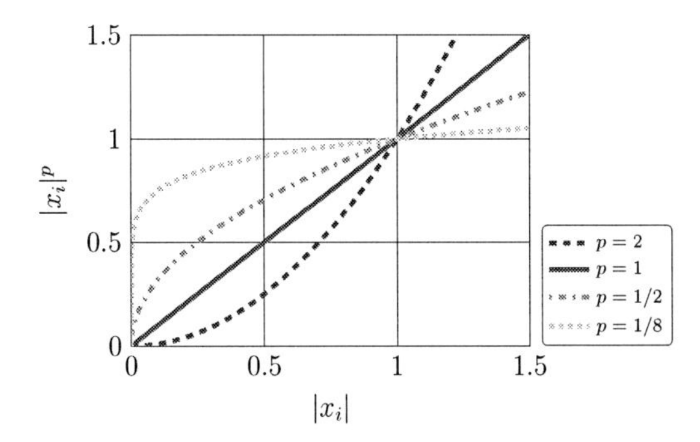

Figure 2.5: Illustration of the influence of p for the penalization of the coefficients $|x_i|$ in the expansion (2.49).

We see that the classical parameter choice rule (2.43) is a sufficient assumption for the regularization property of the Tikhonv-functional with sparsity constraints. The authors of [17] showed under slightly different assumptions on F and the weights w_i also a convergence rate and introduced an algorithm for calculating the minimizer of (2.50). For this propose, they defined a surrogate functional

$$\Gamma_{\alpha,\mathbf{w},p,\delta}(x,a) = T^p_{\alpha,\mathbf{w}}(x,y^\delta) - \|Fx - Fa\|^2 + \|x - a\|^2,$$

where $a \in H_1$ is an arbitrary element. If $\|F\| < 1$ is assumed, the surrogate functional is strictly convex. The main advantage of the surrogate functional is that the nonlinear equations are decoupled because the coupling term $\|Fx\|^2$ cancels out. This can be seen by rearranging the functional as

$$\begin{aligned} \Gamma_{\alpha,\mathbf{w},p,\delta}(x,a) &= \sum_i \left[x_i^2 - 2x_i\left(a - F^*\left(Fa - y^\delta\right)\right)_i + \alpha w_i |x_i|^p\right] \\ &\quad + \|y^\delta\|^2 + \|a\|^2 - \|Fa\|^2. \end{aligned}$$

Since minimizing $\Gamma_{\alpha,\mathbf{w},p,\delta}(\cdot, a)$ is equivalent to minimizing

$$x \mapsto \sum_i x_i^2 - 2x_i\left(a - F^*\left(Fa - y^\delta\right)\right)_i + \alpha w_i |x_i|^p \tag{2.51}$$

with respect to $x \in H_1$, the first order optimality condition for $p > 1$ is given for all i by

$$x_i + \frac{\alpha p w_i}{2}\operatorname{sgn}(x_i)|x_i|^{p-1} = \left(a - F^*\left(Fa - y^\delta\right)\right)_i.$$

For $p = 1$ we have to apply the subdifferential calculus introduced in Section 2.2 in order to determine the optimality condition. By Propositions 16 and 15, we obtain

$$\begin{aligned} & 0 \in \partial\left(x_i^2 - 2x_i\left(a - F^*\left(Fa - y^\delta\right)\right)_i + \alpha w_i |x_i|\right), \\ \Leftrightarrow\ & 0 \in \left(2x_i - 2\left(a - F^*\left(Fa - y^\delta\right)\right)_i + \alpha w_i \partial|x_i|\right), \\ \Leftrightarrow\ & \left(a - F^*\left(Fa - y^\delta\right)\right)_i \in \left(I + \frac{\alpha w_i}{2}\partial|\cdot|\right)x_i, \\ \Leftrightarrow\ & x_i = \left(I + \frac{\alpha w_i}{2}\partial|\cdot|\right)^{-1}\left(a - F^*\left(Fa - y^\delta\right)\right)_i, \\ \Leftrightarrow\ & x_i = \operatorname{sgn}\left(a - F^*\left(Fa - y^\delta\right)\right)_i\left[\left|\left(a - F^*\left(Fa - y^\delta\right)\right)_i\right| - \frac{\alpha w_i}{2}\right]_+, \end{aligned}$$

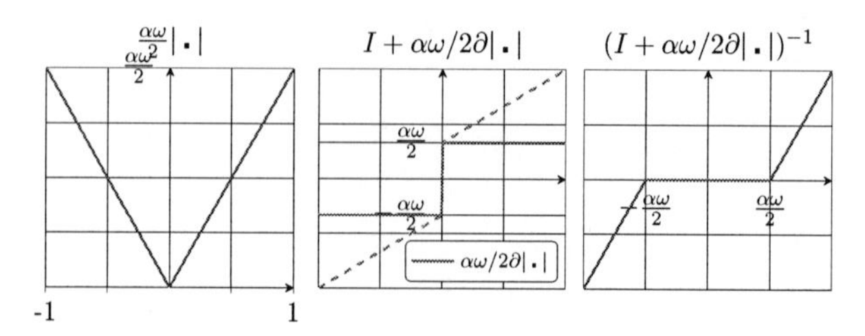

Figure 2.6: Illustration of the computation of the optimality condition for $p = 1$.

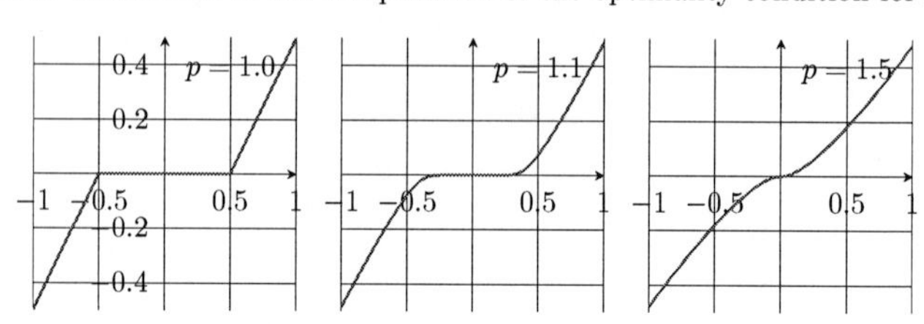

Figure 2.7: Shrinkage-function $S_{\alpha w,p}(x)$ for $\alpha w = 1$ and different p, compare with equation (2.52).

where f_+ denotes the non-negative part of f. Moreover, note that the function $\left(I + \frac{\alpha w_i}{2}\partial\,|\cdot|\right)^{-1}$ is single-valued, see Figure 2.6.

Introducing the shrinkage functions

$$S_{\alpha w,p}(x) = \begin{cases} \operatorname{sgn}(x)\left[|x| - \frac{\alpha w}{2}\right]_+, & p = 1\,, \\ G^{-1}_{\alpha w,p}(x)\,, & 1 < p \leq 2 \end{cases} \tag{2.52}$$

with

$$G_{\alpha w,p}(x) = x + \frac{\alpha w}{2}\operatorname{sgn}(x)\,|x|^{p-1}\,,$$

the soft-shrinkage operator

$$\mathbf{S}_{\alpha\mathbf{w},p}(x) = \sum_i S_{\alpha w,p}(x_i)\varphi_i$$

can be defined. Therefore, the minimizer of (2.51) can be expressed as

$$x = \mathbf{S}_{\alpha\mathbf{w},p}\left(a - F^*\left(Fa - y^\delta\right)\right).$$

It can be shown that the shrinkage functions and operators form contractions on bounded domains, see [17]. In Figure 2.7 the shrinkage function is illustrated for different p. It can be seen that small coefficients are less damped for increasing p.

An iterative minimization algorithm has been proposed in [17] as follows:

$$\begin{aligned} x^0 & \quad \text{arbitrary} \\ x^{n+1} &= \mathbf{S}_{\alpha\mathbf{w},p}\left(x^n - F^*\left(Fx^n - y^\delta\right)\right). \end{aligned}$$

This iterative scheme called *iterated soft shrinkage* comprises at each step a Landweber iteration (compare with (2.40)) followed by a shrinkage. It can be shown that it converges linearly, see [14, Theorem 1].

There are several other minimization algorithms, like the iterated hard thresholding [13] or active set methods like the semi-smooth Newton method (SSN) [24] or the feature sign search algorithm (FSS) [38]. A modification of the latter will be shortly introduced in the following.

For $p = 1 = w_i$ and linear operator F mapping between Hilbert spaces, the Tikhonov-functional (2.50),

$$T_\alpha^1\left(x, y^\delta\right) = \frac{1}{2}\left\|F(x) - y\right\|_2^2 + \alpha \sum_{i\in\mathbb{N}} \left|\langle x \,|\varphi_i\rangle\right| ,$$

can be rewritten as

$$\Psi_\alpha\left(x, y^\delta\right) = \frac{1}{2}\left\|Kx - y\right\|_2^2 + \alpha\left\|x\right\|_{l^1} \tag{2.53}$$

if K is defined as the operator

$$\begin{aligned} K: \quad l^2 \quad &\to H_2 ,\\ \{x_i\}_i \quad &\mapsto F\sum_{i\in\mathbb{N}} x_i\varphi_i = F\sum_{i\in\mathbb{N}} \langle x\,|\varphi_i\rangle\,\varphi_i .\end{aligned}$$

Hereby, $\{\varphi_i\}_i$ is like before an orthonormal basis of H_1 (It could also be an overcomplete dictionary.), but the minimization is done in the space of the coefficients, i.e.

$$\min_{x\in l^2} \Psi_\alpha(x, y^\delta) .$$

Instead of minimizing the l^1- functional Ψ_α, the so called elastic-net functional

$$\Psi_{\alpha,\beta}\left(x, y^\delta\right) = \frac{1}{2}\left\|Kx - y\right\|_2^2 + \alpha\left\|x\right\|_{l^1} + \frac{1}{2}\beta\left\|x\right\|_{l^2}^2 \tag{2.54}$$

for $\alpha, \beta \geq 0$ is introduced. For $\beta > 0$ or K injective this functional is strictly convex and admits thus a unique minimizer. By Proposition 15 and 16, the optimality condition of (2.54) is given by

$$0 \in \partial\Psi_{\alpha,\beta}\left(x, y^\delta\right) = \partial\Psi_\alpha(x) + \beta x$$

which is equivalent to

$$-K^*\left(Kx - y^\delta\right) - \beta x \in \alpha\,\mathrm{sign}\left(x\right)$$

(see Example 9 and 13). Using the shrinkage function (2.52), this condition can be rewritten as

$$S_\alpha\left(-K^*\left(Kx - y^\delta\right)\right) = \beta x$$

if $\beta > 0$. In [28] the stability of the minimizer of the elastic net functional has been proven. In order to introduce the modified FSS-algorithm, the term of consistency is needed.

Definition 43. Let $K : \mathbb{R}^s \to \mathbb{R}^m$, $y^\delta \in \mathbb{R}^m$ and $\bar{s} = \{1, 2, \ldots, s\}$. For $A \subset \bar{s}$, $x = (x_i)_{i\in\bar{s}} \in \mathbb{R}^s$ and $\theta = \{\theta_i\}_{i\in\bar{s}} \in \{-1, 0, 1\}^s$, a triple (A, x, θ) is called *consistent* if

$$\begin{aligned} i \in A \quad &\Longrightarrow \quad \mathrm{sign}(x_i) = \theta_i \neq 0 ,\\ i \in A^c \quad &\Longrightarrow \quad x_i = \theta_i = 0 .\end{aligned}$$

Due to the a consistent triple, the optimality condition can be divided into

$$\left(-K^*\left(Kx-y^\delta\right)-\beta x\right)_i \;=\alpha\theta_i\,,\;\; i\in A\,, \tag{2.55}$$

$$\left|K^*\left(Kx-y^\delta\right)\right|_i \;\leq\alpha\,,\;\; i\in A^c\,. \tag{2.56}$$

This formulation allows to consider sub-problems of the form

$$\Psi_{A\theta}=\frac{1}{2}\left\|K_Ax_A-y^\delta\right\|^2+\alpha\left\langle x_A\,|\theta_A\right\rangle,$$

where K_A denotes the matrix containing only the columns of K with index in the active set A. This fact is used in the regularized feature sign search algorithm 2.1 (RFSS). This algorithm will be used in Chapter 4 for solving the parameter identification problem. A detailed description of the algorithm and numerical examples can be found in [54].

The next example illustrates the difference between classical Tikhonov-regularization with L_2-penalty and the sparsity promoting regularization with L_1penalty.

Example 44 (Differentiation example 2)**.** The integral operator of Example 32 resp. 38 is again considered. The real solution x is sparse in the Haar-wavelet basis, only 15 coefficients are non-zero, see the solid line in Figure 2.8a. The corresponding data computed by $y(t)=\int_0^t x(s)$ ds as well as a noised version are plotted in Figure 2.8b.

Approximations of the derivatives of the noisy data are computed by minimization of the Tikhonov-functional (2.53) with L_1-penalty by using the RFSS-algorithm. The computed solutions x_α^δ are shown for different regularization parameters in Figure 2.8a. Like expected the solution for the biggest parameter $\alpha=10^{-1}$ is the sparsest one with only 6 non-zero coefficients. Obviously, this parameter is to big. For smaller regularization parameter the solutions become less sparse but better approximations for the true solution.

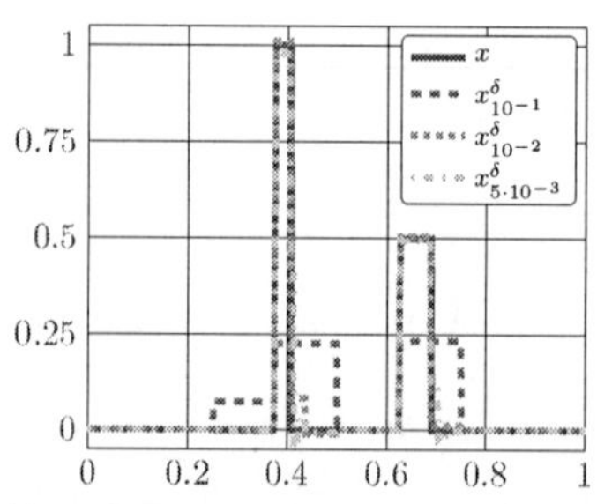

(a) Real sparse solution x as well as computed regularized solutions x_α^δ by minimizing the functional (2.53).

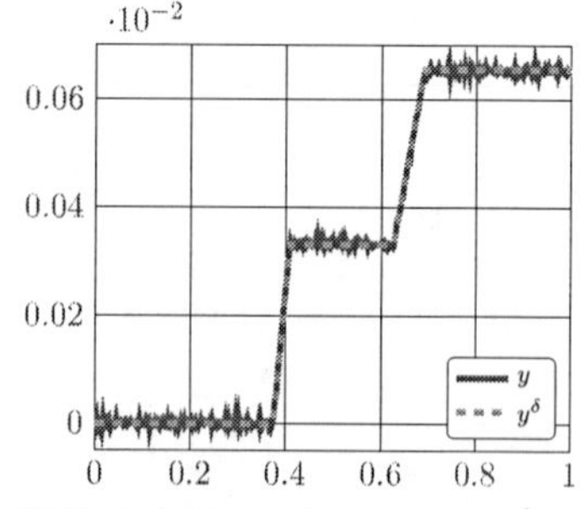

(b) Real data y and noisy data y^δ with $\left\|y-y^\delta\right\|<0.01$.

Figure 2.8: Differentiation example 2: Real sparse solution as well as computed approximations x_α^δ and real data y as well as given noisy data y^δ.

Algorithmus 2.1 Regularized feature sign search algorithm [28].

Initialization: Consistent (A_0, x^0, θ^0), e.g. $A_0 = \emptyset$, $x^0 = 0$, $\theta^0 = 0$

Step 1: Check the optimality condition:

if fulfilled **then** $\{x_0$ is optimal$\}$
  return solution x^0
else if (2.56) is violated **then**
  go to Step 2
else
  go to Step 3
end if

Step 2: Increasing the active set

Take index $i_0^k = \arg\min_{i \in A_{k-1}^c} \left|K^* \left(Kx - y^\delta\right)\right|_i - \alpha$, violating (2.56) most

$A_k = A_{k-1} \cup \{i_0^k\}$

* $$\theta_i^k = \begin{cases} \theta_i^{k-1}, & i \neq i_0^k, \\ -\mathrm{sign}\left(K^* \left(Kx^{k-1} - y^\delta\right)\right)_{i_0^k}, & i = i_0^k. \end{cases}$$

Step 3: Optimization

Compute x^k such that (2.55) is fulfilled, i.e.

$x^k|_{A_k} = (\beta + M_{A_k})^{-1} \left(K^* y^\delta - \alpha\theta^k\right)|_{A_k}$

$x^k|_{A_k^c} = 0$

if $\left(A_k, x^k, \theta^k\right)$ consistent **then**
  go to Step 5
else
  go to Step 4
end if

Step 4: Decreasing the active set

Compute smallest $\lambda_0 \in (0,1)$ such that $\left(\lambda_0 x^k + (1-\lambda_0)x^{k-1}\right)$ changes the sign

Denote the corresponding index with i_0

$x^k = \lambda_0 x^k + (1-\lambda_0)x^{k-1}$

$A_k = A_k \setminus i_0$ $\{$remove index i_0 from active set$\}$

$\theta_{i_0}^k = 0$

if (2.56) holds **then**
  go to Step 5
else
  $A_{k+1} = A_k, \quad \theta^{k+1} = \theta^k$
  $k = k + 1$
  go to Step 3
end if

Step 5: Optimality check

if (2.56) holds **then**
  return x^k
else
  $k = k + 1$
  go to Step 2
end if

In Figure 2.9 the L_1-regularization method is compared with the classical L_2-norm regularization. The latter produces non-sparse but smooth solutions in contrast to the solution of the functional with sparsity constraints. Note that the regularization parameter is fixed for both methods in the same way in order to compare the solutions. Consequently, it may not be the best choice for both methods. In particular, the parameter is still chosen not small enough in order to obtain the best solution for the L_2-regularization, see Figure 2.9b and 2.9d. On the other side, the choice of $\alpha = 10^{-3}$ is already chosen to small for the sparsity approach as it can be seen in Figure 2.9c.

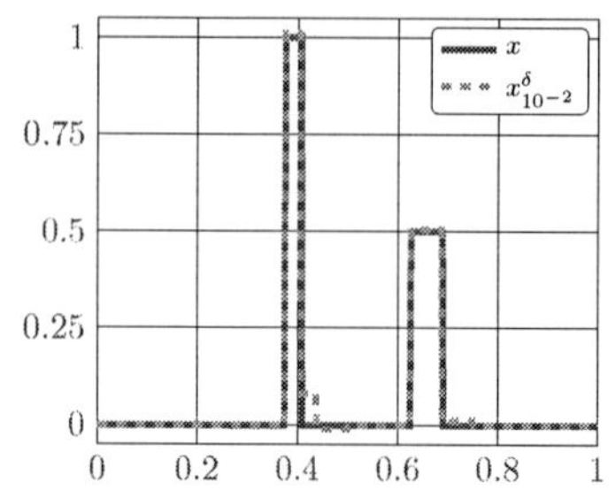

(a) Solution x^δ_α of (2.53) for $\alpha = 10^{-2}$. This value seems to be near the optimal regularization parameter.

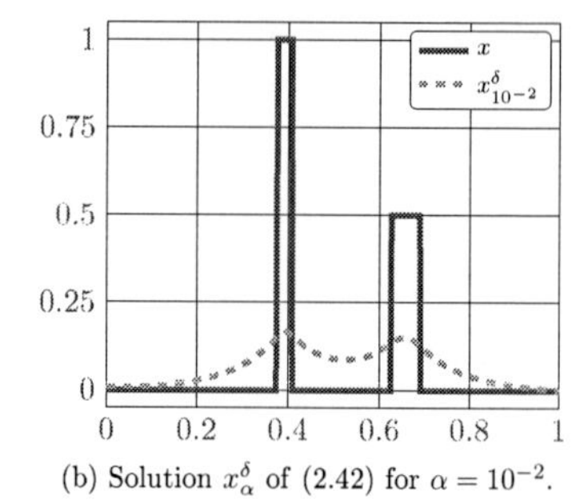

(b) Solution x^δ_α of (2.42) for $\alpha = 10^{-2}$.

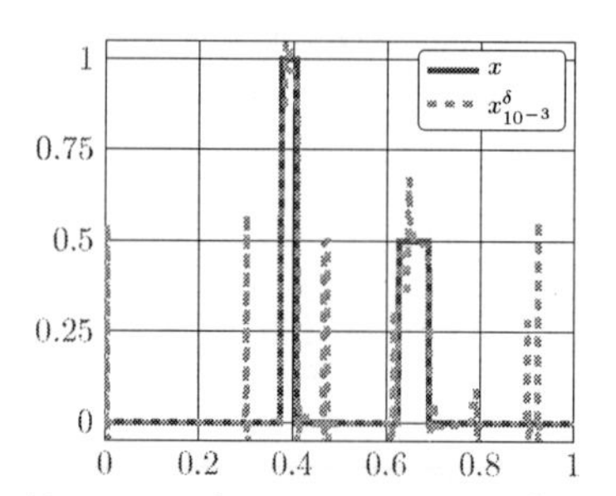

(c) Solution x^δ_α of (2.53) for $\alpha = 10^{-3}$. For a better reconstruction, a bigger α should be chosen.

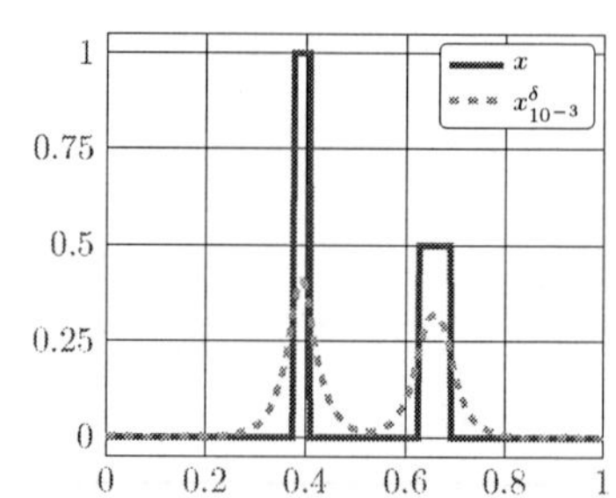

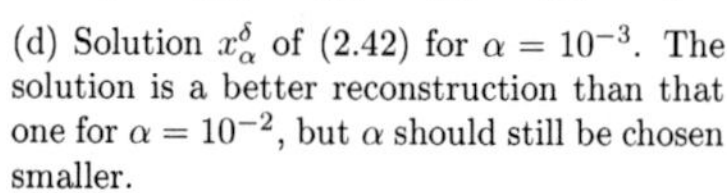

(d) Solution x^δ_α of (2.42) for $\alpha = 10^{-3}$. The solution is a better reconstruction than that one for $\alpha = 10^{-2}$, but α should still be chosen smaller.

Figure 2.9: Comparison between L_1-regularization (left hand side) and classical Tikhonov-regularization with L_2-penalty (right hand side). The L_1-regularization problem (2.53) has been solved with the RFSS-algorithm, the classical Tikhonv-functional (2.42) with help of the filter function (see Remark 37). The example shows that L_1-regularization produces sparse solutions, whereas the classical L_2-solutions are smooth. The regularization parameter is fixed for both approaches to the same values in order to compare the different kind of solutions. As a consequence, the parameter is not optimally chosen for both methods.

The differentiation example illustrates that L_1-regularization indeed promotes sparse solutions. In Chapter 4 the nonlinear parameter identification problem will be reformulated in a way that two linear operator equations have to be solved where one is the integral operator. Therefore, several additional examples for computed sparse derivatives are treated in Subsection 4.2.

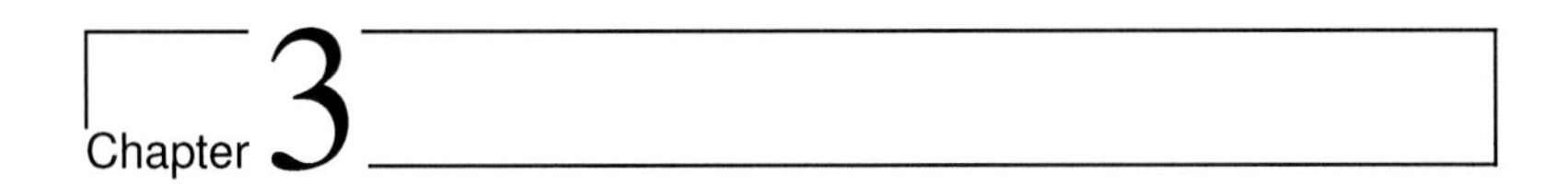

A forward model describing ultra precision turning operations

3.1 Introduction

In this chapter a forward model for an ultra precision turning lathe, including the process machine interaction, is presented. As explained in the introducing Chapter 1, ultra precision machining is in use to manufacture optical components or mechanical parts which have high requirements regarding the surface quality. In order to achieve the desired surface qualities, the understanding of the surface generation process and its main influencing parameters is necessary. One of the most important parameters is the residual unbalance of the main spindle because unbalance driven oscillations can lead to a significant decrease of surface quality [67]. Therefore, the minimization of the residual unbalance is one of the main objectives in diamond machining, and consequently the balancing process of the spindle has to be done accurately.

Balancing means the mounting of additional weights at certain positions at the rotating parts of the machine in such a way that the machine vibrations measured at different sensor positions are reduced below a certain tolerance. In order to predict the surface quality of the machined surface, a model is necessary. The model must include not only the influences of the unbalances but also the process machine interaction because the cutting process may excite additional vibrations or amplify unbalance driven ones which may lead to a poor surface quality. Such a model which describes the interaction between unbalances occurring during the machining process and the engine-shaft structure is presented in the next section. As an exemplary cutting process, ultra precision turning is considered. The interaction model is based on two sub-models like illustrated in Figure 3.1.

First, a process model for the simulation of the process parameters and the resulting cutting forces is developed in Subsection 3.2.1. Second, a machine or structure model for the turning lathe, which computes the vibrations for given unbalances, is built in Subsection 3.2.2. Both sub-models are coupled in Subsection 3.2.3 into a mechanic-dynamical model where the impact of unbalance vibrations on the movement of the workpiece and the tool is modeled and vice versa. Output of the interaction model are the displacements of the workpiece and the tool as well as the process parameters and therefore the relative position of the tool to the workpiece. This information is used in the material removal model for the computation

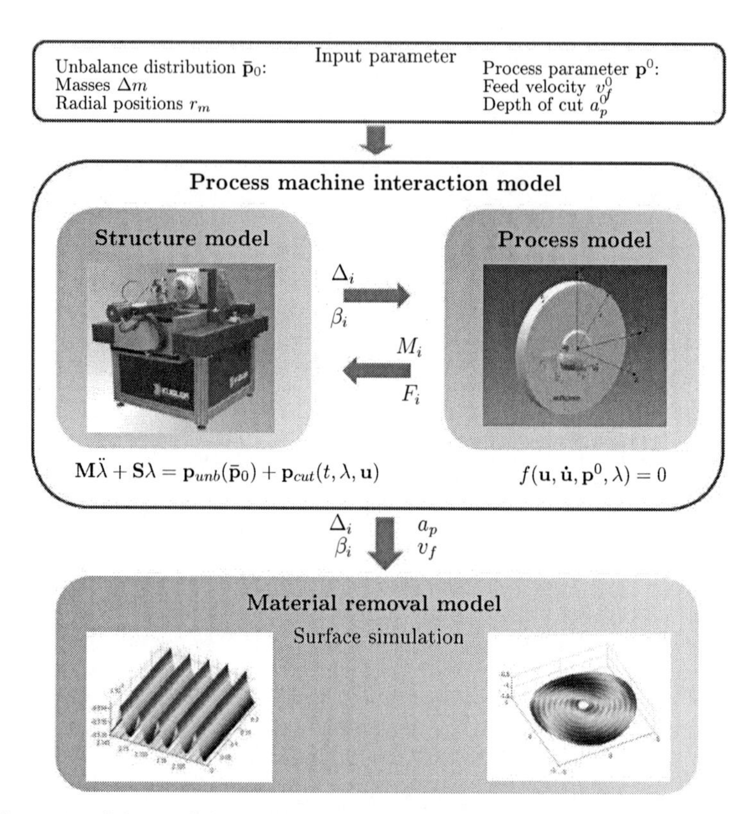

Figure 3.1: Scheme of the process machine interaction model.
The interaction model comprises a structure model in order to simulate the machine vibrations λ and a process model which computes the process forces F_i and the real process parameters like depth of cut a_p and the feed velocity v_f. Since the process forces act on the workpiece, they induce moments M_i and unbalances to the machine structure, and therefore forces and moments are input parameters for the structure model. On the other hand, the vibrating machine structure influences the process itself such that the displacements Δ and tilting angles β of the workpiece have to be considered in the process model. Finally, the displacements of workpiece and tool as well as the process kinematics determine the relative tool workpiece position and thus the resulting surface topography which is simulated with help of the material removal model. Consequently, output of the interaction model is the global form and the micro-topography of the machined surface.

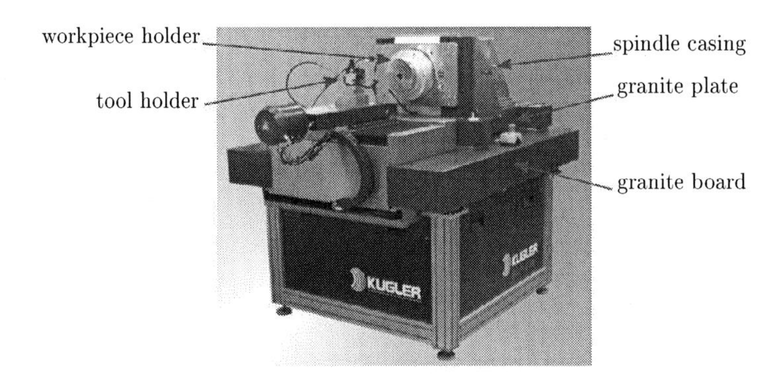

Figure 3.2: Test stand for ultra precision turning operations at the Laboratory for Precision Machining, see for example [36].

of the material which is cut by the tool. This model provides thus the possibility to simulate the resulting surface and hence to determine the achievable surface quality of the workpiece, which will be shown in Section 3.3.

The ultra precision turning lathe utilized for the experiments is not a conventional lathe but a test stand, which is in particular designed for a research project within the framework of the priority program SPP1180 of the German Research Foundation in order to investigate the influences of unbalances on the surface quality. The test stand shown in Figure 3.2 has been set up by the Laboratory for Precision Machining (LFM). It offers the possibility to use dual-plane-balancing capabilities. Conversely, up to now spindle balancing in conventional diamond machining has been done usually via single-plane balancing. For a detailed description of the test stand and the balancing capabilities as well as for the experimental setup for the force measurements, we refer to [6, 7, 36].

The interaction model has been built in collaboration with the Radon Institute of Computational and Applied Mathematics (RICAM) at Linz and the LFM at Bremen. In particular, all experiments in this chapter were conducted by Andreas Krause at the LFM, and the machine model in Subsection 3.2.2 has been developed by Jenny Niebsch at the RICAM.

3.2 The process-machine interaction model

In this section a model for the process machine interaction of the ultra precision turning lathe, introduced in the previous introduction of this chapter, is presented being used in the following chapter for the determination of optimal process parameters given desired surface profiles.

3.2.1 The process model for face turning

Basis of the process model are three main ideas. First, a force model is needed which computes the process forces in dependence of the relevant process parameters. The second idea consist of the fact that the forces act on the tool and thus cause elastic deflections of the tool holder. The deflections are determined by the forces via a proportional relation. The third ingredient of the process model is the idea that the deflections caused by the forces influence the process parameters because they cause displacements of the tool and therefore of the tool position relative to the workpiece. This influence of the elastic deflections is modeled in the model for the tool path and the actual process parameters. Actual parameter means that the parameters is influenced by the elastic deflections and vibrations in contrast to the given fixed input parameters at the machine.

Figure 3.3 shows a schematic diagram of the considered diamond face turning process. In face turning, the tool is moving along the x-axis with a feed velocity v_f and cuts the workpiece with a depth of cut a_p at its front face. The acting force can be split in three components, the cutting force F_c in negative y-direction, the passive force F_p in z-direction (sometimes also denoted as thrust force F_t), and the feed force F_f in negative x-direction. In the considered process, the cutting velocity $v_c(t) = 2\pi n(r - d(t))$ is decreasing with time because the rotational speed n is constant, but the traveled distance d is increasing. Therefore, a force model is needed which includes the cutting velocity v_c in addition to the depth of cut a_p and the feed rate f defined as the distance which the diamond tool is traveling during one revolution.

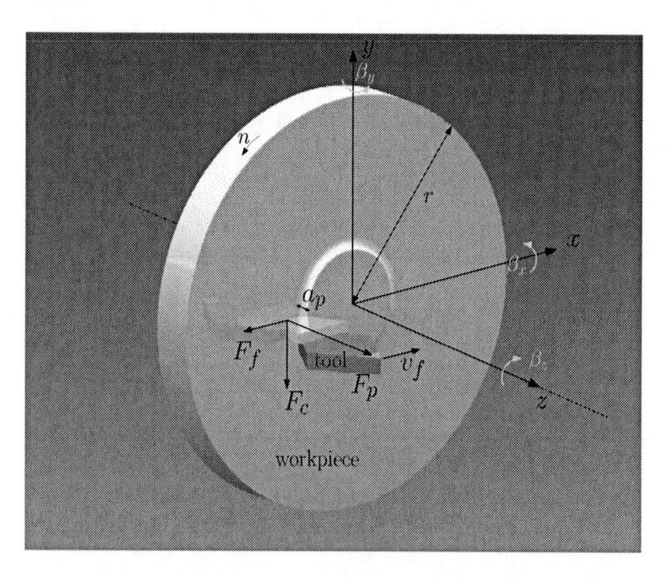

Figure 3.3: Scheme of the face turning process.

In the next subsection a new force model for ultra precision turning is developed. Afterward, the influence of the elastic deflections on the actual parameters is modeled in Subsection 3.2.1.2. Finally, in Subsection 3.2.1.3 several simplifying assumptions on the force model are made in order to obtain a mathematical less complex process model, which can be used in Chapter 4 for the mathematical analysis of the forward model. Each simplifying assumption of the force model results

Table 3.1: Experimental conditions for diamond turning experiments.

parameter		value range	standard condition	
rational spindle speed	n	800...1500 rev/min	n_0	1500 rev/min
feed velocity	v_f	4...12 mm/min	f_0	0.008 mm/rev
depth of cut	a_p	2...14 μm	$a_{p,0}$	0.005 mm

fixed parameter for all experiments		
tool		mono-crystalline diamond, radius tool
tool nose radius	r_ϵ	760 μm
workpiece radius	r	30 mm
workpiece material		AlMg3

in a simplified process model. Consequently, a hierarchy of force models with decreasing accuracy and related process models with decreasing complexity is built and summarized in Table 3.6 at the end of section.

3.2.1.1 A force model for ultra precision turning

The idea of the force model (1.4) for micro turning, presented in the introductory chapter, is adapted in order to elaborate a new force model for ultra precision face turning. The normalized specific cutting force is thus represented as a product of functions g_i depending on the depth of cut a_p, the feed rate f, the rotational speed n and the cutting velocity v_c, i.e.

$$k_c/k_{c0} = g_1(a_p)g_2(f)g_3(v_c)g_4(n)\,. \tag{3.1}$$

Here, functions g_1 and g_3 will have the same form as f_1 and f_2 in the model (1.4). Similar equations hold for the specific passive and feed forces k_p and k_f. The product structure of the approach can be interpreted as several correcting factors, similar to the constant correcting factors introduced in the classical Kienzle law for macro cutting processes. In contrast to the model (1.4), the dependence of the friction and the cutting edge radius is not considered because both parameters has been kept constant in the experiments. However, if necessary, their influence could be modeled by simply adding further functions depending on these parameters.

The form of the functions g_i $(i = 1, 2, 3, 4)$ and the model constants therein are determined by a least square fit to force measurements for different cutting conditions, see Table 3.1 for details, followed by a averaging over the different series of measurements. Therefore, several experiments have been performed, where only one parameter has been varied while the remaining cutting parameters have been fixed to their standard value. The standard values are: $n_0 = 1500$ min^{-1}, $f_0 = 0.005$ mm, $a_{p,0} = 0.005$ mm, and $v_{c,0} = 4320$ mm/s. Since the normalized specific forces and therefore the functions g_i equal one for standard conditions, the constants of each functions g_i can be determined separately. In Table 3.2 the determined function parameters are summarized.

The measurements of the specific forces over the depth of cut and feed rate are illustrated in Figure 3.4 together with the fitted curves g_1 and g_2. The measurements suggest the for both function being of the form of function f_1, i.e.

$$g_{1,i}(a_p) = c_{a,i}a_p^{-m_{c,i}} \qquad (i = f, c, p) \tag{3.2}$$

Table 3.2: Model constants for the force model.

	Passive force $i = p$	Cutting force $i = c$	Feed force $i = f$
$g_{1,i}(a_p)$	$c_{a,p} = 0.04$ $m_{c,p} = 0.63$	$c_{a,c} = 0.26$ $m_{c,c} = 0.26$	$c_{a,f} = 0.001$ $m_{c,f} = 1.00$
$g_{2,i}(f)$	$c_{f,p} = 0.01$ $m_{d,p} = 1.00$	$c_{f,c} = 0.03$ $m_{d,c} = 0.70$	$c_{f,f} = 0.002$ $m_{d,f} = 0.89$
$g_{3,i}(v_c)$	$\alpha_{1,p} = 0.64$ $\alpha_{2,p} = 0.10$ $\beta_{1,p} = 0.05$ $\beta_{2,p} = 8.60$	$\alpha_{1,c} = 0.59$ $\alpha_{2,c} = 0.10$ $\beta_{1,c} = 0.06$ $\beta_{2,c} = 9.78$	$\alpha_{1,f} = 0.17$ $\alpha_{2,f} = 0.10$ $\beta_{1,f} = 0.09$ $\beta_{2,f} = 16.76$
$g_{4,i}(n)$	$\gamma_{1,p} = 0.85\,10^{-6}$ $\gamma_{2,p} = 5.09\,10^{6}$ $\kappa_{1,p} = 1.60$ $\kappa_{2,p} = 2.87$	$\gamma_{1,c} = 2.73\,10^{-5}$ $\gamma_{2,c} = 5.09\,10^{6}$ $\kappa_{1,c} = 1.43$ $\kappa_{2,c} = 2.39$	$\gamma_{1,f} = 1.86\,10^{-3}$ $\gamma_{2,f} = 4.82\,10^{6}$ $\kappa_{1,f} = 1.17$ $\kappa_{2,f} = 46.52$
$k_{i,0}\left[\frac{10^3\text{N}}{\text{mm}^2}\right]$	$k_{p,0} = 19.98$	$k_{c,0} = 3.59$	$k_{f,0} = 0.07$

and

$$g_{2,i}(f) = c_{f,i} f^{-m_{d,i}} \qquad (i = f, c, p)\,. \tag{3.3}$$

Note that the dependence on a_p is the similar to power law of Kienzle, compare with (1.2).

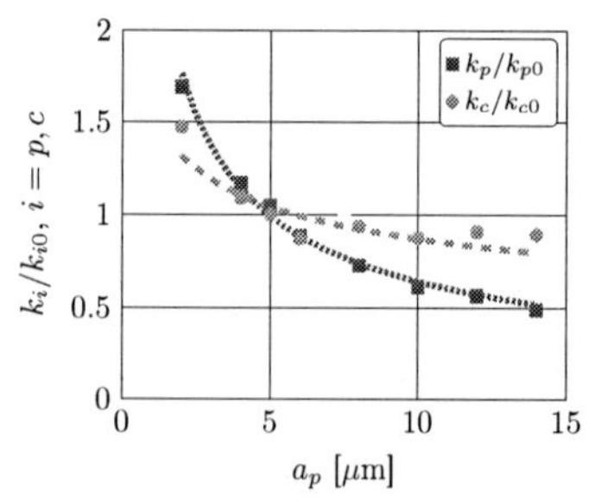

(a) Measured normalized specific force and function g_1 over depth of cut.

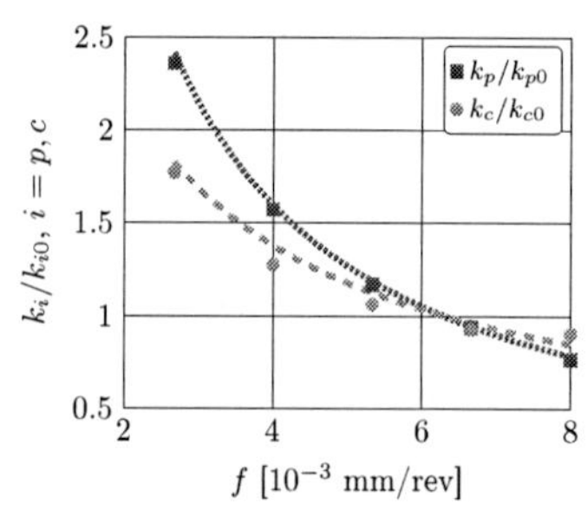

(b) Measured normalized specific force and function g_2 over feed.

Figure 3.4: The measured normalized specific passive and cutting forces are represented by squares and circles. The fitted curves g_1 and g_2 are plotted as dotted and dashed lines for the passive and cutting force component.

The function f_2 in (1.4) is taken as an approach to model the function g_3, i.e.

$$g_{3,i}(v_c) = \alpha_{1,i} v_c^{\beta_{1,i}} + \alpha_{2,i} v_c^{-\beta_{2,i}} \qquad (i = f, c, p)\,. \tag{3.4}$$

The function g_4 is determined to be of the same form

$$g_{4,i}(n) = \gamma_{1,i} n^{\kappa_{1,i}} + \gamma_{2,i} n^{-\kappa_{2,i}} \qquad (i = f, c, p). \tag{3.5}$$

The functions g_3 and g_4 as well as the measurements of the specific forces over cutting velocity and rotational spindle speed are shown in Figure 3.5.

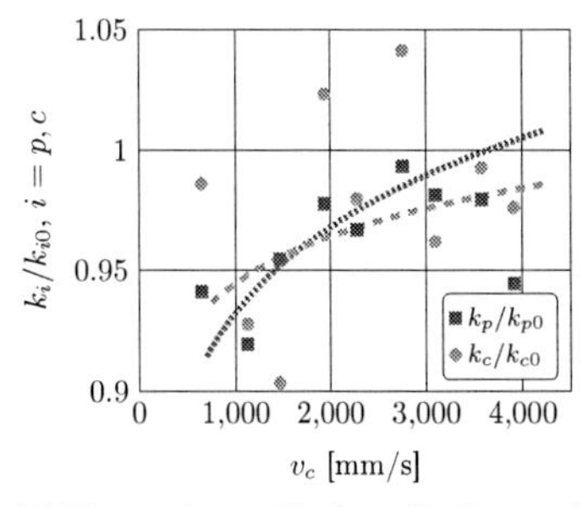

(a) Measured normalized specific force and function g_3 over cutting velocity.

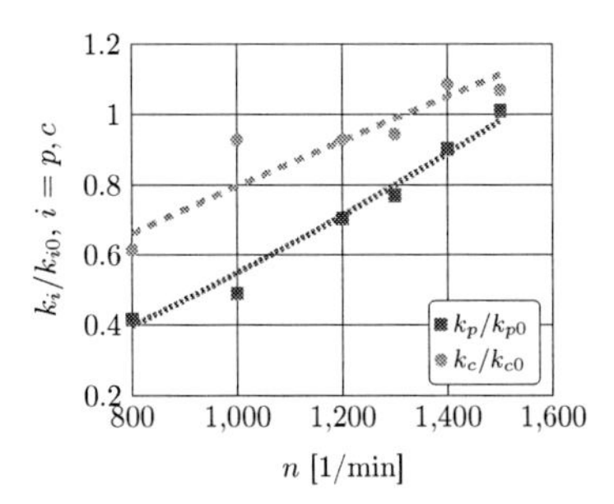

(b) Measured normalized specific force and function g_4 over rotational speed n.

Figure 3.5: The measured normalized specific passive and cutting forces are represented by squares and circles. The fitted curves g_3 and g_4 are plotted as dotted and dashed lines for the passive and cutting force component.

Overall the specific force k_i for $i = f, c, p$ is thus given by

$$k_i = (k_{i0} c_{a,i} c_{f,i})\, a_p^{-m_{c,i}} f^{-m_{d,i}} \left(\alpha_{1,i} v_c^{\beta_{1,i}} + \alpha_{2,i} v_c^{-\beta_{2,i}}\right) \left(\gamma_{1,i} n^{\kappa_{1,i}} + \gamma_{2,i} n_i^{-\kappa_{2,i}}\right). \tag{3.6}$$

Using the specific cutting forces, we are now able to calculate the cutting force components via the usual relationship

$$F_c = k_c A_c, \qquad F_p = k_p A_c, \quad F_f = k_f A_c \tag{3.7}$$

for the forces by Kienzle, see [59]. Here, A_c denotes the cross sectional area of cut shown in Figure 3.6 which can be approximated by $A_c \approx a_p f$, see [49, Section 2.2.5]. Therefore, the force component F_i $(i = f, c, p)$ is given by

$$F_i = \bar{k}_{i0}\, a_p^{1-m_{c,i}} f^{1-m_{d,i}} \left(\alpha_{1,i} v_c^{\beta_{1,i}} + \alpha_{2,i} v_c^{-\beta_{2,i}}\right) \left(\gamma_{1,i} n^{\kappa_{1,i}} + \gamma_{2,i} n_i^{-\kappa_{2,i}}\right) \tag{3.8}$$

with $\bar{k}_{i0} := (k_{i0} c_{a,i} c_{f,i})$.

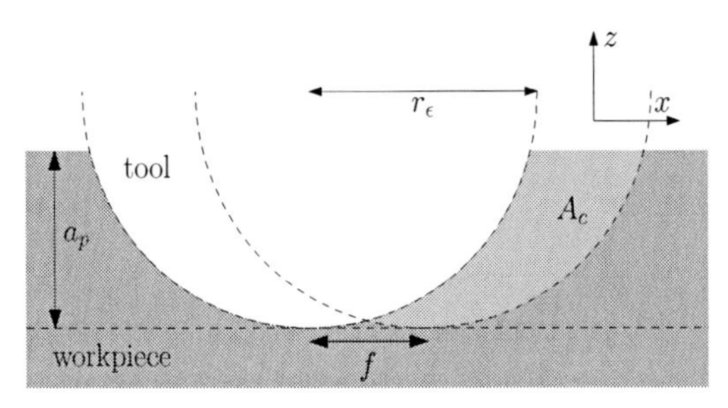

Figure 3.6: Illustration of the cross-sectional area of cut A_c.

This force model is used in the following in order to compute the displacements of the tool caused by elastic deflections and the actual process parameters. Note that the product structure of the model allows to advance it by adding further functions of other relevant parameters if necessary without the need to determine all model constants again. Moreover, it may also possible to extend the model by adding a second force term in analogy to the approach (1.1) of Altintas for macro cutting processes in order to incorporate the influence of friction on the cutting edge, but therefore further experimental investigations are necessary. Another possible refinement of the model would be a more precise computation of the cross-sectional area of cut A_c, either analytical as e.g. in [34] or numerically with help of the surface simulation model presented in Section 3.3. This approach would also provide a back coupling to the force model.

In Subsection 3.2.1.3, the force model is again investigated and simplified in order to obtain simplified resulting process model. The result is a chain of force model with decreasing complexity, see Table 3.6 at the end of this section.

3.2.1.2 Simulation of the tool path

The second part of the process model comprises the simulation of the actual process parameters and the tool path. Main idea of the model is that the actual tool tip position is not only given by the kinematics of the tool, but it is influenced by displacements of the tool holder and of the workpiece. In particular, a deflected tool holder causes a reduced depth of cut like illustrated in Figure 3.7. Ignoring the workpiece vibrations for instance, the description of the tool position on the surface is given by the movement of the tool and its displacements δ_i $(i = x, y, z)$ in the direction i, i.e. by

$$\begin{aligned} x(t) &= -r + \int_0^t v_f^0(s)\,\mathrm{d}s && -\delta_x(t), && (3.9)\\ y(t) &= && -\delta_y(t), && (3.10)\\ z(t) &= -a_p^0(t) + \tfrac{1}{2l_h}\left(\delta_x^2(t) + \delta_y^2(t)\right) && +\delta_z(t), && (3.11) \end{aligned}$$

where l_h denotes the length of the tool holder. With a_p^0 and v_f^0 the fixed input depth of cut and feed velocity are denoted, i.e. the depth of cut and feed velocity which are given as inputs at the machine. Figure 3.7b shows an illustration for the computation of the expression for the reduced depth of cut in (3.11), which has been first published in [45].

If in addition the vibrations of the workpiece are included, i.e. the displacements Δ_i $(i = x, y, z)$ and the rotational tilt β_i $(i = x, y, z)$, the tool path description is extended by

$$\begin{aligned} x(t) &= -r + \int_0^t v_f^0(s)\,\mathrm{d}s && -\delta_x(t) - \Delta_x(t)\,, && (3.12)\\ y(t) &= && -\delta_y(t) - \Delta_y(t)\,, && (3.13)\\ z(t) &= -a_p^0(t) + \tfrac{1}{2l_h}\left(\delta_x^2(t) + \delta_y^2(t)\right) && +\delta_z(t) - \Delta_z(t) + x(t)\tan(\beta_y(t)). && (3.14) \end{aligned}$$

The angle β_y denotes the angle caused by a rotation of the workpiece around the y-axis. This tilt of the workpiece can influence the actual depth of the cut, which is described by the last term in equation (3.14).

Since the deflections affect the actual tool position and therefore the process parameters, we introduce new time-dependent process parameters, namely the actual depth of cut

$$a_p(t) = -z(t) \qquad (3.15)$$

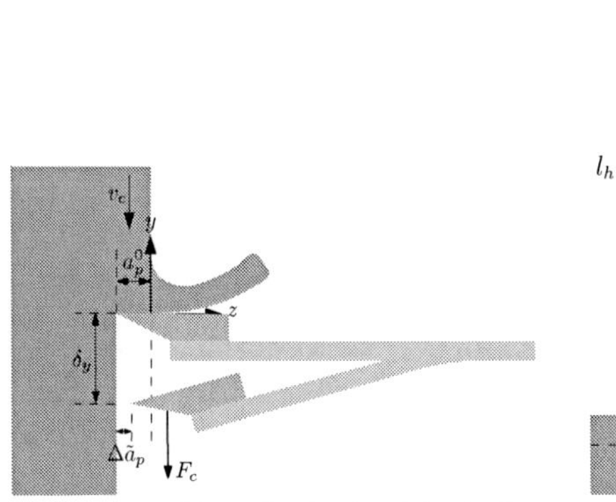

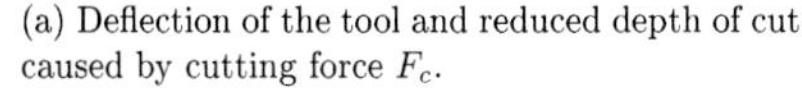
(a) Deflection of the tool and reduced depth of cut caused by cutting force F_c.

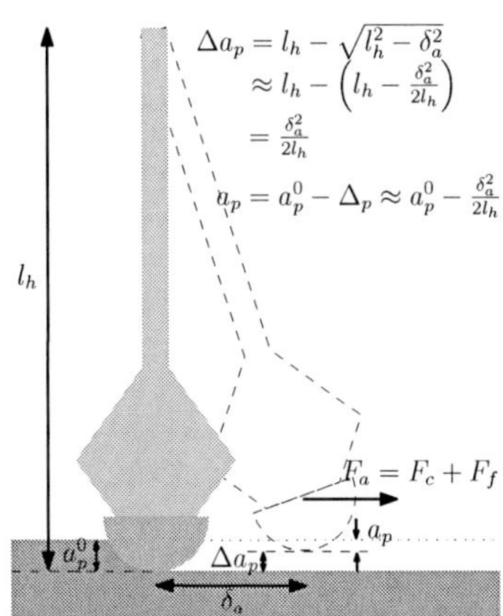

(b) Idealized deflection $\delta_a = \sqrt{\delta_x^2 + \delta_y^2}$ caused by active force $F_a = F_f + F_c$.

Figure 3.7: Illustration of a deflected tool holder and the resulting reduction of the depth of cut.

defined as the negative actual position z and the actual feed velocity

$$v_x(t) = \dot{x}(t) \tag{3.16}$$

given as the actual relative velocity of the tool on the workpiece surface. The actual feed rate f is therefore given by $f(t) = n^{-1}v_x(t)$. The actual cutting velocity can be expressed by

$$v_c(t) = -2\pi n\, x(t)\,. \tag{3.17}$$

In contrast to the fixed input parameters of the machine, these new process parameters are called "actual parameters". Note that also the given input parameters at the machine my be time-dependent, but only the actual parameters include the variation caused by elastic deflections of the holder and vibrations of the machine structure.

These actual process parameters are plugged into the equation (3.8) in order to compute the actual forces. Since the rotational speed n is kept constant during the turning process, the function $g_{4,i}(n) = (\gamma_{1,i}n^{\kappa_{1,i}} + \gamma_{2,i}n_i^{-\kappa_{2,i}}) \equiv G_{n,i}$ is constant, too. Therefore, the actual force F_i $(i = f, c, p)$ in terms of $a_p(t)$, $v_x(t)$ and $v_c(t)$ is given by

$$F_i = \bar{F}_{i,n}\, a_p(t)^{1-m_{c,i}} \left(\frac{v_x(t)}{n}\right)^{1-m_{d,i}} \left(\alpha_{1,i}v_c(t)^{\beta_{1,i}} + \alpha_{2,i}v_c(t)^{-\beta_{2,i}}\right), \tag{3.18}$$

where all constants are collected in $\bar{F}_{i,n} := \bar{k}_{i0}G_{n,i} = (k_{i0}c_{a,i}c_{f,i}G_{n,i})$.

Since the actual forces are proportional to the corresponding deflections, the deflections of the tool holder are determined by

$$\delta_x(t) = \frac{F_f(t)}{k_{ex}}\,, \quad \delta_y(t) = \frac{F_c(t)}{k_{ey}}\,, \quad \delta_z(t) = \frac{F_t(t)}{k_{ez}}\,. \tag{3.19}$$

Here k_{ei} denotes the corresponding stiffness in the direction j ($j = x, y, z$). Since all force components have the same structure, we get for all three spatial directions $j = x, y, z$ the deflection

$$\begin{aligned}\delta_j(t) &= \left(\frac{\bar{F}_{i,n}}{k_{ej}} n^{-(1-m_{d,i})}\right) a_p(t)^{1-m_{c,i}} v_x(t)^{1-m_{d,i}} \left(\alpha_{1,i} v_c(t)^{\beta_{1,i}} + \alpha_{2,i} v_c(t)^{-\beta_{2,i}}\right),\\ &= D_j\, a_p(t)^{1-m_{c,i}} v_x(t)^{1-m_{d,i}} \left(\alpha_{1,i} v_c(t)^{\beta_{1,i}} + \alpha_{2,i} v_c(t)^{-\beta_{2,i}}\right) \end{aligned} \tag{3.20}$$

with $D_j := \bar{F}_{i,n} k_{ej}^{-1} n^{m_{d,i}-1}$, ($j = x, y, z$) and ($i = f, c, p$).

In order to simplify the notations, a second notation with index $j = x, y, z$ instead of $j = f, c, p$ indicating the direction of the force component is introduced for all model constants of the force model collected in Table 3.2. Consequently, each model constant has a double notation, for example $m_{c,p} = m_{c,z}$, $\alpha_{1,c} = \alpha_{1,y}$ and $\beta_{2,f} = \beta_{2,x}$. Therefore, only one index $i = j = x, y, z$ appears in (3.20).

To calculate the actual deflections, the actual feed velocity $v_x = \dot{x}$ has to be known. Therefore, the position x must be differentiated which itself depends on the derivative of δ_x. That is why the differentiation of all equations for the positions (3.12)-(3.14) and the deflections (3.19) is necessary in order to derive a system of differential equations. Differentiation of the deflection δ_i ($i = x, y, z$) yields

$$\begin{aligned}\dot{\delta}_i(t) =& D_i \Big\{\left[\alpha_{1,i} v_c(t)^{\beta_{1,i}} + \alpha_{2,i} v_c(t)^{-\beta_{2,i}}\right] \left[(1-m_{c,i})\, a_p(t)^{-m_{c,i}} \dot{a}_p(t) v_x(t)^{1-m_{d,i}}\right.\\ &\left. + (1-m_{d,i})\, v_x(t)^{-m_{d,i}} \dot{v}_x(t) a_p(t)^{1-m_{c,i}}\right]\\ &+ \left[\alpha_{1,i}\beta_{1,i} v_c(t)^{\beta_{1,i}-1} - \alpha_{2,i}\beta_{2,i} v_c(t)^{-\beta_{2,i}-1}\right] \dot{v}_c(t) a_p(t)^{1-m_{c,i}} v_x(t)^{1-m_{d,i}}\Big\},\\ =& D_i a_p(t)^{-m_{c,i}} v_x(t)^{-m_{d,i}} \Big\{\left[\alpha_{1,i} v_c(t)^{\beta_{1,i}} + \alpha_{2,i} v_c(t)^{-\beta_{2,i}}\right]\\ &\left[(1-m_{c,i})\, \dot{a}_p(t) v_x(t) + (1-m_{d,i})\, \dot{v}_x(t) a_p(t)\right]\\ &+ \left[\alpha_{1,i}\beta_{1,i} v_c(t)^{\beta_{1,i}-1} - \alpha_{2,i}\beta_{2,i} v_c(t)^{-\beta_{2,i}-1}\right] \dot{v}_c(t) a_p(t) v_x(t)\Big\}. \end{aligned} \tag{3.21}$$

From the equations for the positions it is ensured that

$$\dot{x}(t) = +v_f^0(t) - \dot{\delta}_x(t) - \dot{\Delta}_x(t)\,, \tag{3.22}$$

$$\dot{y}(t) = -\dot{\delta}_y(t) - \dot{\Delta}_y(t)\,, \tag{3.23}$$

$$\begin{aligned}\dot{z}(t) =& -\dot{a}_p^0(t) + \tfrac{1}{l_h}\left(\delta_x(t)\dot{\delta}_x(t) + \delta_y(t)\dot{\delta}_y(t)\right) + \dot{\delta}_z(t) - \dot{\Delta}_z(t)\\ &+ \dot{x}(t)\tan\left(\beta_y(t)\right) + x(t)\tfrac{\dot{\beta}_y(t)}{\cos^2(\beta_y(t))}\,. \end{aligned} \tag{3.24}$$

Using the relation $\dot{x} = v_x$ and the derivatives of the expression (3.15), i.e

$$\dot{a}_p(t) = -\dot{z}(t)\,,$$

the differential equations

$$\dot{\delta}_x(t) = v_f^0(t) - v_x(t) - \dot{\Delta}_x(t) \tag{3.25}$$

and

$$\begin{aligned}\dot{a}_p(t) =& \dot{a}_p^0(t) - \frac{1}{l_h}\left(\delta_x(t)\dot{\delta}_x(t) + \delta_y(t)\dot{\delta}_y(t)\right) - \dot{\delta}_z(t) + \dot{\Delta}_z(t)\\ &- v_x(t)\tan\left(\beta_y(t)\right) - x(t)\frac{\dot{\beta}_y(t)}{\cos^2\left(\beta_y(t)\right)}\end{aligned}$$

can be derived from (3.22) and (3.24).

The two equations for $\dot{\delta}_x$ can be equalized by

$$\begin{aligned}
\dot{\delta}_x \underset{(3.21)}{=} & D_x a_p^{-m_{c,f}} v_x^{-m_{d,f}} \Big\{ \dot{v}_c a_p v_x \left[\alpha_{1,f}\beta_{1,f} v_c^{\beta_{1,f}-1} - \alpha_{2,f}\beta_{2,f} v_c^{-\beta_{2,f}-1} \right] \\
& + \left[\alpha_{1,f} v_c^{\beta_{1,f}} + \alpha_{2,f} v_c^{-\beta_{2,f}} \right] \left[(1-m_{c,f})\, \dot{a}_p v_x + (1-m_{d,f})\, \dot{v}_x a_p \right] \Big\}, \\
\underset{(3.25)}{=} & v_f^0 - v_x - \dot{\Delta}_x \,,
\end{aligned}$$

where the time dependency is omitted in order to simplify the notation. Solving this equation for $\dot{v}_x$, the differential equation

$$\begin{aligned}
\dot{v}_x = & \frac{D_x^{-1}\left[v_f^0 - v_x - \dot{\Delta}_x\right] a_p^{m_{c,f}} v_x^{m_{d,f}} - \left[\alpha_{1,f} v_c^{\beta_{1,f}} + \alpha_{2,f} v_c^{-\beta_{2,f}} \right] (1-m_{c,f})\, \dot{a}_p v_x}{\left[\alpha_{1,f} v_c^{\beta_{1,f}} + \alpha_{2,f} v_c^{-\beta_{2,f}} \right] (1-m_{d,f})\, a_p} \\
& - \frac{\dot{v}_c a_p v_x \left[\alpha_{1,f}\beta_{1,f} v_c^{\beta_{1,f}-1} - \alpha_{2,f}\beta_{2,f} v_c^{-\beta_{2,f}-1} \right]}{\left[\alpha_{1,f} v_c^{\beta_{1,f}} + \alpha_{2,f} v_c^{-\beta_{2,f}} \right] (1-m_{d,f})\, a_p}, \\
= & \frac{D_x^{-1}\left[v_f^0 - v_x - \dot{\Delta}_x\right] a_p^{m_{c,f}-1} v_x^{m_{d,f}}}{\left[\alpha_{1,f} v_c^{\beta_{1,f}} + \alpha_{2,f} v_c^{-\beta_{2,f}} \right] (1-m_{d,f})} - \frac{(1-m_{c,f})}{(1-m_{d,f})} \frac{\dot{a}_p}{a_p} v_x \\
& + \frac{2\pi n v_x^2 \left[\alpha_{1,f}\beta_{1,f} v_c^{\beta_{1,f}-1} - \alpha_{2,f}\beta_{2,f} v_c^{-\beta_{2,f}-1} \right]}{\left[\alpha_{1,f} v_c^{\beta_{1,f}} + \alpha_{2,f} v_c^{-\beta_{2,f}} \right] (1-m_{d,f})}
\end{aligned} \tag{3.26}$$

is derived, where in the last line the expression

$$\dot{v}_c(t) = -2\pi n \dot{x}(t) = -2\pi n v_x(t) \tag{3.27}$$

is applied. Summarizing the equations (3.22)-(3.27) yields a system of ordinary differential equations

$$\begin{aligned}
\dot{v}_c(t) &= -2\pi n v_x(t)\,, \\
\dot{a}_p(t) &= \dot{a}_p^0(t) - \frac{1}{l_h}\Big(\delta_x(t)\dot{\delta}_x(t) + \delta_y(t)\dot{\delta}_y(t)\Big) - \dot{\delta}_z(t) + \dot{\Delta}_z(t) \\
&\qquad - v_x(t)\tan\left(\beta_y(t)\right) + \frac{v_c(t)}{2\pi n}\frac{\dot{\beta}_y(t)}{\cos^2\left(\beta_y(t)\right)}\,,
\end{aligned} \tag{3.28}$$

$$\begin{aligned}
\dot{v}_x(t) = & \frac{2\pi n v_x(t)^2 \left[\alpha_{1,x}\beta_{1,x} v_c(t)^{\beta_{1,x}-1} - \alpha_{2,x}\beta_{2,x} v_c(t)^{-\beta_{2,x}-1}\right]}{(1-m_{d,x})\left[\alpha_{1,x} v_c(t)^{\beta_{1,x}} + \alpha_{2,x} v_c(t)^{-\beta_{2,x}}\right]} \\
& + \frac{D_x^{-1}\dot{\delta}_x(t) a_p(t)^{m_{c,x}-1} v_x(t)^{m_{d,x}}}{(1-m_{d,x})\left[\alpha_{1,x} v_c(t)^{\beta_{1,x}} + \alpha_{2,x} v_c(t)^{-\beta_{2,x}}\right]} \\
& - \frac{(1-m_{c,x})}{(1-m_{d,x})}\frac{\dot{a}_p(t)}{a_p(t)} v_x(t)\,,
\end{aligned} \tag{3.29}$$

$$\dot{\delta}_x(t) = v_f^0(t) - v_x(t) - \dot{\Delta}_x(t)\,, \tag{3.30}$$

$$\begin{aligned}\dot{\delta}_y(t) = D_y a_p(t)^{-m_{c,y}} v_x(t)^{-m_{d,y}} &\left\{\left[\alpha_{1,y} v_c(t)^{\beta_{1,y}} + \alpha_{2,y} v_c(t)^{-\beta_{2,y}}\right]\right.\\ &\cdot\left[(1-m_{c,y})\,\dot{a}_p(t) v_x(t) + (1-m_{d,y})\,\dot{v}_x(t) a_p(t)\right]\\ &\left.+\left[\alpha_{1,y}\beta_{1,y} v_c(t)^{\beta_{1,y}-1} - \alpha_{2,y}\beta_{2,y} v_c(t)^{-\beta_{2,y}-1}\right]\dot{v}_c(t) a_p(t) v_x(t)\right\},\end{aligned} \tag{3.31}$$

$$\begin{aligned}\dot{\delta}_z(t) = D_z a_p(t)^{-m_{c,z}} v_x(t)^{-m_{d,z}} &\left\{\left[\alpha_{1,z} v_c(t)^{\beta_{1,z}} + \alpha_{2,z} v_c(t)^{-\beta_{2,z}}\right]\right.\\ &\cdot\left[(1-m_{c,z})\,\dot{a}_p(t) v_x(t) + (1-m_{d,z})\,\dot{v}_x(t) a_p(t)\right]\\ &\left.+\left[\alpha_{1,z}\beta_{1,z} v_c(t)^{\beta_{1,z}-1} - \alpha_{2,z}\beta_{2,z} v_c(t)^{-\beta_{2,z}-1}\right]\dot{v}_c(t) a_p(t) v_x(t)\right\}.\end{aligned} \tag{3.32}$$

This is a implicitly given differential equation system of the form

$$\dot{\mathbf{u}} = f(\mathbf{u}, \dot{\mathbf{u}}, \mathbf{p}^0, \mathbf{q}^0) \tag{3.33}$$

with

$$\mathbf{u}(t) = (v_c(t), a_p(t), v_x(t), \delta_x(t), \delta_y(t), \delta_z(t)),$$

where the right hand side f depends on the given input parameters and the vibrations of the workpiece collected in the vectors

$$\mathbf{p}^0(t) = \left(\dot{a}_p^0(t), v_f^0(t)\right) \text{ and } \mathbf{q}^0(t) = \left(\dot{\Delta}_x(t), \dot{\Delta}_y(t), \dot{\Delta}_z(t), \beta_y(t), \dot{\beta}_y(t)\right). \tag{3.34}$$

Introducing the abbreviations b_i, $(i = 1, \ldots, 11)$, defined in Table 3.3, the differential system can be rewritten as

$$\begin{aligned}\dot{u}_1(t) =& -b_1\, u_3(t),\\ \dot{u}_2(t) =& p_1^0(t) - b_{11}\left(u_4(t)\dot{u}_4(t) + u_5(t)\dot{u}_5(t)\right) - \dot{u}_6(t) + q_3(t)\\ & - u_3(t)\tan\left(q_4(t)\right) + \frac{u_1(t)}{b_1}\frac{q_5(t)}{\cos^2\left(q_4(t)\right)},\\ \dot{u}_3(t) =& \frac{b_1 u_3(t)^2\left[b_{12} u_1(t)^{\beta_{1,x}-1} - b_{13} u_1(t)^{-\beta_{2,x}-1}\right]}{b_5\left[\alpha_{1,x} u_1(t)^{\beta_{1,x}} + \alpha_{2,x} u_1(t)^{-\beta_{2,x}}\right]}\\ & + \frac{b_2^{-1}\dot{u}_4(t) u_2(t)^{-b_8} u_3(t)^{1-b_5}}{b_5\left[\alpha_{1,x} u_1(t)^{\beta_{1,x}} + \alpha_{2,x} u_1(t)^{-\beta_{2,x}}\right]} - \frac{b_8\,\dot{u}_2(t)}{b_5\, u_2(t)} u_3(t),\\ \dot{u}_4(t) =& p_2^0(t) - u_3(t) - q_1(t),\\ \dot{u}_5(t) =& b_3 u_2(t)^{b_9-1} u_3(t)^{b_6-1}\left\{\left[\alpha_{1,y} u_1(t)^{\beta_{1,y}} + \alpha_{2,y} u_1(t)^{-\beta_{2,y}}\right]\right.\\ &\cdot\left[b_9\dot{u}_2(t) u_3(t) + b_6\dot{u}_3(t) u_2(t)\right]\\ &\left.- b_1\left[b_{14} u_1(t)^{\beta_{1,y}-1} - b_{15} u_1(t)^{-\beta_{2,y}-1}\right] u_2(t) u_3^2(t)\right\},\\ \dot{u}_6(t) =& b_4 u_2(t)^{b_{10}-1} u_3(t)^{b_7-1}\left\{\left[\alpha_{1,z} u_1(t)^{\beta_{1,z}} + \alpha_{2,z} u_1(t)^{-\beta_{2,z}}\right]\right.\\ &\cdot\left[b_{10}\dot{u}_2(t) u_3(t) + b_7\dot{u}_3(t) u_2(t)\right]\\ &\left.- b_1\left[b_{16} u_1(t)^{\beta_{1,z}-1} - b_{17} u_1(t)^{-\beta_{2,z}-1}\right] u_2(t) u_3^2(t)\right\}.\end{aligned} \tag{ODE 1}$$

We will call this differential equation system complete process model in the following.

In order to solve the differential equation system, initial conditions for $\mathbf{u}$ and $\dot{\mathbf{u}}$ are necessary. Starting point for their derivation are equations (3.12) and (3.14) for the positions, (3.20) for the deflections and the expressions (3.15), (3.17) and (3.16) for the actual process parameters, all evaluated at the point $t_0 = 0$, i.e.

$$\begin{aligned}x(0) &= -r - \delta_x(0) - \Delta_x(0) =: x_0,\\ z(0) &= -a_p^0(0) + \frac{1}{2} b_{11}\left(\delta_x^2(0) + \delta_y^2(0)\right) + \delta_z(0) - \Delta_z(0) + x_0\tan(\beta_y(0)),\\ \delta_i(0) &= D_i\, a_p^{1-m_{c,i}}(0) v_x^{1-m_{d,i}}(0)\left(\alpha_{1,i} v_c(0)^{\beta_{1,i}} + \alpha_{2,i} v_c(0)^{-\beta_{2,i}}\right), i = x, y, z\end{aligned}$$

Table 3.3: Definitions of the model constants and abbreviations.

Definition of b_i		Definition c_i	
b_1	$2\pi n$	c_1	$-r-\Delta_x(0)$
b_2	D_x	c_2	$D_x\, a_{p,0}^{b_8} v_{x,0}^{b_5}$
b_3	D_y	c_3	$D_y\, a_{p,0}^{b_9} v_{x,0}^{b_6}$
b_4	D_z	c_4	$D_z\, a_{p,0}^{b_{10}} v_{x,0}^{b_7}$
b_5	$1-m_{d,x}$	c_5	$p_1^0(0)+\dot{\Delta}_z(0)-v_{x,0}\tan\left(\beta_y(0)\right)$
b_6	$1-m_{d,y}$	c_6	$\frac{\dot{\beta}_y(0)}{b_1\cos^2(\beta_y(0))}$
b_7	$1-m_{d,z}$	c_7	$b_1 v_{x,0}^2$
b_8	$1-m_{c,x}$	c_8	$b_2^{-1} a_{p,0}^{-b_8} v_{x,0}^{1-b_5}$
b_9	$1-m_{c,y}$	c_9	$\frac{b_8}{b_5}\frac{v_{x,0}}{a_{p,0}}$
b_{10}	$1-m_{c,z}$	c_{10}	$p_2^0(0)-v_{x,0}-\dot{\Delta}_x(0)$
b_{11}	$1/l_h$	c_{11}	$b_3 a_{p,0}^{b_9-1} v_{x,0}^{b_6-1}$
b_{12}	$\alpha_{1,x}\beta_{1,x}$	c_{12}	$b_1 a_{p,0} v_{x,0}^2$
b_{13}	$\alpha_{2,x}\beta_{2,x}$	c_{13}	$b_4 a_{p,0}^{b_{10}-1} v_{x,0}^{b_7-1}$
b_{14}	$\alpha_{1,y}\beta_{1,y}$		
b_{15}	$\alpha_{2,y}\beta_{2,y}$		
b_{16}	$\alpha_{1,z}\beta_{1,z}$		
b_{17}	$\alpha_{2,z}\beta_{2,z}$		

as well as $a_p(0)=-z(0)$, $v_c(0)=-b_1x(0)$, and $v_y(0)=\dot{x}(0)$. The initial conditions for $\dot{\mathbf{u}}$ are given by the differential equations systems (3.28) - (3.32) evaluated at $t=0$. For chosen values

$$u_2(0)=a_p^0(0)=:a_{p,0}$$

and

$$u_3(0)=v_x(0)=:v_{x,0}\,,$$

follows that

$$\dot{u}_1(0)=\dot{v}_c(0)=-b_1v_{x,0}$$

and

$$\begin{aligned}
x(0) &= c_1-u_4(0)\,,\\
-u_6(0) &= \frac{1}{2}b_{11}\left(u_4^2(0)+u_5^2(0)\right)-\Delta_z(0)+x(0)\tan(\beta_y(0))\,,\\
u_4(0) &= c_2\left(\alpha_{1,i}u_1(0)^{\beta_{1,i}}+\alpha_{2,i}u_1(0)^{-\beta_{2,i}}\right),\\
u_5(0) &= c_3\left(\alpha_{1,i}u_1(0)^{\beta_{1,i}}+\alpha_{2,i}u_1(0)^{-\beta_{2,i}}\right),\\
u_1(0) &= -b_1x(0)\,,\\
\dot{u}_2(0) &= c_5-b_{11}\left(u_4(0)\dot{u}_4(0)+u_5(0)\dot{u}_5(0)\right)-\dot{u}_6(0)\\
&\quad +c_6u_1(0)\,,\\
\dot{u}_3(0) &= \frac{c_7\left[\alpha_{1,f}\beta_{1,f}u_1(0)^{\beta_{1,f}-1}-\alpha_{2,f}\beta_{2,f}u_1(0)^{-\beta_{2,f}-1}\right]}{b_5\left[\alpha_{1,f}u_1(0)^{\beta_{1,f}}+\alpha_{2,f}u_1(0)^{-\beta_{2,f}}\right]}\\
&\quad \frac{c_8\dot{u}_4(0)}{b_5\left[\alpha_{1,f}u_1(0)^{\beta_{1,f}}+\alpha_{2,f}u_1(0)^{-\beta_{2,f}}\right]}-c_9\dot{u}_{2,}(0)\,,
\end{aligned}$$

$$\begin{aligned}
\dot{u}_4(0) &= c_{10}\,,\\
\dot{u}_5(0) &= -c_{11}\left\{\left[\alpha_{1,y}u_1(0)^{\beta_{1,y}}+\alpha_{2,y}u_1(0)^{-\beta_{2,y}}\right]\right.\\
&\qquad\cdot\left[b_9\dot{u}_2(0)v_{x,0}+b_6\dot{u}_3(0)a_{p,0}\right]\\
&\quad\left.-c_{12}\left[\alpha_{1,y}\beta_{1,y}u_1(0)^{\beta_{1,y}-1}-\alpha_{2,y}\beta_{2,y}u_1(0)^{-\beta_{2,y}-1}\right]\right\},\\
\dot{u}_6(0) &= -c_{13}\left\{\left[\alpha_{1,z}u_1(0)^{\beta_{1,z}}+\alpha_{2,z}u_1(0)^{-\beta_{2,z}}\right]\right.\\
&\qquad\cdot\left[b_{10}\dot{u}_2(0)v_{x,0}+b_7\dot{u}_3(0)a_{p,0}\right]\\
&\quad\left.-c_{12}\left[\alpha_{1,z}\beta_{1,z}u_1(0)^{\beta_{1,z}-1}-\alpha_{2,z}\beta_{2,z}u_1(0)^{-\beta_{2,z}-1}\right]\right\},
\end{aligned}$$

where the abbreviations c_i defined in Table 3.3 are used. This system is solved numerically such that we obtain for the initial conditions

$$\begin{aligned}
\mathbf{u}_0 &= \left(u_1(0),a_{p,0},v_{x,0},u_4(0),u_5(0),u_6(0)\right)\,,\\
\dot{\mathbf{u}}_0 &= \left(-b_1v_{x,0},\dot{u}_2(0),\dot{v}_x(0),\dot{u}_4(0),\dot{u}_5(0),\dot{u}_6(0)\right)\,.
\end{aligned}$$

Another possibility is to fix as starting point in $x-$direction

$$x(0)=-r$$

the edge of the workpiece such that follows

$$\begin{aligned}
u_4(0) &= c_1+r=-\Delta_x(0) &&=:\delta_{x,0}\,,\\
u_1(0) &= b_1r &&=:v_{c,0}\,,\\
u_5(0) &= c_3\left(\alpha_{1,y}{v_{c,0}}^{\beta_{1,y}}+\alpha_{2,y}{v_{c,0}}^{-\beta_{2,y}}\right) &&=:\delta_{y,0}\,,\\
u_6(0) &= \tfrac{1}{2}b_{11}\left(\delta_{x,0}^2+\delta_{y,0}^2\right)-\Delta_z(0)-r\tan\left(\beta_y(0)\right) &&=:\delta_{z,0}\,.
\end{aligned}$$

Together with $\dot{u}_4(0)=c_{10}$ the remaining equations

$$\begin{aligned}
\dot{u}_2(0) &= c_5-b_{11}\left(c_{10}\delta_{x,0}+\delta_{y,0}\dot{u}_5(0)\right)-\dot{u}_6(0)+c_6v_{c,0}\,,\\
\dot{u}_3(0) &= \frac{c_8c_{10}+c_7\left[\alpha_{1,f}\beta_{1,f}{v_{c,0}}^{\beta_{1,f}-1}-\alpha_{2,f}\beta_{2,f}{v_{c,0}}^{-\beta_{2,f}-1}\right]}{b_5\left[\alpha_{1,f}{v_{c,0}}^{\beta_{1,f}}+\alpha_{2,f}{v_{c,0}}^{-\beta_{2,f}}\right]}-c_9\dot{u}_2(0)\,,\\
\dot{u}_5(0) &= c_{11}\left\{\left[\alpha_{1,y}{v_{c,0}}^{\beta_{1,y}}+\alpha_{2,y}{v_{c,0}}^{-\beta_{2,y}}\right]\left[b_9\dot{u}_2(0)v_{x,0}+b_6\dot{u}_3(0)a_{p,0}\right]\right.\\
&\quad\left.-c_{12}\left[\alpha_{1,y}\beta_{1,y}{v_{c,0}}^{\beta_{1,y}-1}-\alpha_{2,y}\beta_{2,y}{v_{c,0}}^{-\beta_{2,y}-1}\right]\right\}\,,\\
\dot{u}_6(0) &= -c_{13}\left\{\left[\alpha_{1,z}{v_{c,0}}^{\beta_{1,z}}+\alpha_{2,z}{v_{c,0}}^{-\beta_{2,z}}\right]\left[b_{10}\dot{u}_2(0)v_{x,0}+b_7\dot{u}_3(0)a_{p,0}\right]\right.\\
&\quad\left.-c_{12}\left[\alpha_{1,z}\beta_{1,z}{v_{c,0}}^{\beta_{1,z}-1}-\alpha_{2,z}\beta_{2,z}{v_{c,0}}^{-\beta_{2,z}-1}\right]\right\}\,,
\end{aligned}$$

can be solved explicitly, namely

$$\begin{aligned}
\dot{u}_2(0) &= c_{19}/\left(c_{20}+1\right)\\
\dot{u}_3(0) &= c_{14}-c_9c_{19}/\left(c_{20}+1\right)\,,\\
\dot{u}_5(0) &= c_{16}+c_{15}c_{19}/\left(c_{20}+1\right)\,,\\
\dot{u}_6(0) &= c_{18}+c_{17}c_{19}/\left(c_{20}+1\right)\,,
\end{aligned}$$

where the short notations

$$c_{14}:=\frac{c_8c_{10}+c_7\left[\alpha_{1,f}\beta_{1,f}{v_{c,0}}^{\beta_{1,f}-1}-\alpha_{2,f}\beta_{2,f}{v_{c,0}}^{-\beta_{2,f}-1}\right]}{b_5\left[\alpha_{1,f}{v_{c,0}}^{\beta_{1,f}}+\alpha_{2,f}{v_{c,0}}^{-\beta_{2,f}}\right]}\,,$$

$$
\begin{aligned}
c_{15} &:= c_{11}\left[\alpha_{1,y}{v_{c,0}}^{\beta_{1,y}} + \alpha_{2,y}{v_{c,0}}^{-\beta_{2,y}}\right]\left[b_9 v_{x,0} + b_6 c_9 a_{p,0}\right],\\
c_{16} &:= c_{11}\left(\left[\alpha_{1,y}{v_{c,0}}^{\beta_{1,y}} + \alpha_{2,y}{v_{c,0}}^{-\beta_{2,y}}\right]\left[b_6 c_{14} a_{p,0}\right]\right.\\
&\quad \left. -c_{12}\left[\alpha_{1,y}\beta_{1,y}{v_{c,0}}^{\beta_{1,y}-1} - \alpha_{2,y}\beta_{2,y}{v_{c,0}}^{-\beta_{2,y}-1}\right]\right),\\
c_{17} &:= c_{13}\left[\alpha_{1,z}{v_{c,0}}^{\beta_{1,z}} + \alpha_{2,z}{v_{c,0}}^{-\beta_{2,z}}\right]\left[b_{10} v_{x,0} + b_7 c_9 a_{p,0}\right],\\
c_{18} &:= c_{13}\left(\left[\alpha_{1,z}{v_{c,0}}^{\beta_{1,z}} + \alpha_{2,z}{v_{c,0}}^{-\beta_{2,z}}\right]\left[b_{11} c_{15} a_{p,0}\right]\right.\\
&\quad \left. -c_{12}\left[\alpha_{1,z}\beta_{1,z}{v_{c,0}}^{\beta_{1,z}-1} - \alpha_{2,z}\beta_{2,z}{v_{c,0}}^{-\beta_{2,z}-1}\right]\right),\\
c_{19} &:= c_5 - b_{11}\left(c_{10}\delta_{x,0} + \delta_{y,0}c_{16}\right) - c_{18} + c_6 v_{c,0},\\
c_{20} &:= b_{11}c_{15}\delta_{y0} + c_{17}
\end{aligned}
$$

are introduced. A numerical example will be treated in Subsection 3.2.3.

In the numerical test examples, it turned out that the equation system is numerically difficult to handle because it is a stiff problem. One crucial point is that a starting feed velocity $v_{x,0} = 0$ would be desirable because then the deflections would be zero at the begin of the process, too. However, this is not possible because the specific force is not defined at this point. Since in the equation system terms are appearing which are proportional to $v_x^{-m_d}$, also a choice of $v_{x,0}$ near 0 can lead to wrong results i.e. too big forces and hence non consistent positions (i.e. if plugging the results into the position equation (3.12)-(3.14), they are not fulfilled.). Therefore, we will use in Chapter 4 a modified equation system which assumes the ansatz function $g_2(f)$ to be constant what corresponds to $m_{d,i} = 0$, $i = x, y, z$. Note that nevertheless the forces still depend on f respectively v_x in a linear way. With this choice, it is also possible to choose $v_{x,0} = 0$ as initial condition for the actual feed velocity.

3.2.1.3 Simplifications of the process model

In order to facilitate the analysis of the process model in the next chapter and to accelerate its numerical solution of the coupled interaction model, simplifications of the force model will be investigated in this subsection.

In the Subsection 3.2.1.1 the formula (3.6) for the specific forces $k_{f,}$, k_c and k_p has been derived. Since all force components show a similar behavior, the assumption is made that the specific forces k_f and k_c can be determined from k_p by a proportional relation, i.e. by $k_f = c_f k_p$ and $k_c = c_c k_p$. Such kind of assumption is often made for force models, see for example [59] for the case of the Kienzle-model for conventional cutting processes, where the feed and passive force components are assumed to be proportional to the cutting force. Here, the proportional relationship is developed in dependence of the passive force because it is the dominant component in the measurements. More precisely, for each function g_i $(i = 1, 2, 3, 4)$ shall hold that

$$g_{1,x}(a_p) = c_{a,x}g_1(a_p), \qquad g_{1,y}(a_p) = c_{a,y}g_1(a_p), \tag{3.35}$$
$$g_{2,x}(f\,) = c_{f,x}g_2(f), \qquad g_{2,y}(f\,) = c_{f,y}g_2(f), \tag{3.36}$$
$$g_{3,x}(v_c) = c_{v,x}g_3(v_c), \qquad g_{3,y}(v_c) = c_{v,y}g_3(v_c), \tag{3.37}$$
$$g_{4,x}(n\,) = c_{n,x}g_4(n), \qquad g_{4,y}(n\,) = c_{n,y}g_4(n), \tag{3.38}$$

where the index z for indication of the z-direction is suppressed. The proportional factors are determined by a least square fit to the force measurements, see Table 3.4. The approximated functions $g_{j,x}$ and $g_{j,y}$ $(j = 1, 2, 3, 4)$ as well as the functions $g_j := g_{j,z} (j = 1, 2, 3, 4)$ are shown in Figure 3.8 and Figure 3.9.

Table 3.4: Proportional factors for (3.35)-(3.38)

	Cutting force $i = y$	Feed force $i = x$
$g_{1,i}$	$c_{a,y} = 0.98$	$c_{a,x} = 1.56$
$g_{2,i}$	$c_{f,y} = 0.81$	$c_{f,x} = 0.36$
$g_{3,i}$	$c_{v,y} = 0.92$	$c_{v,x} = 0.35$
$g_{41,i}$	$c_{n,y} = 1.22$	$c_{n,x} = 2.35$

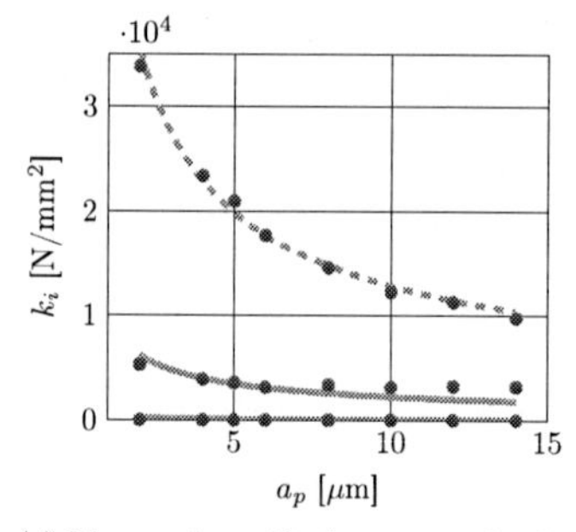

(a) Measured specific forces over depth of cut and simplified model functions $g_{1,x}$ and $g_{1,y}$ given by (3.35).

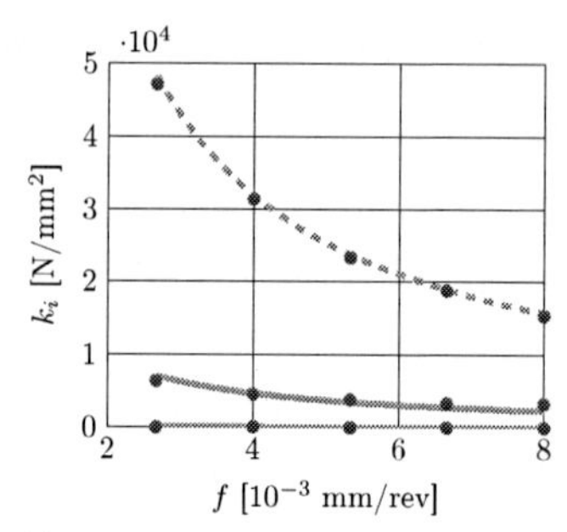

(b) Measured specific forces over feed and simplified model functions $g_{2,x}$ and $g_{2,y}$ given by (3.36).

Figure 3.8: The measured specific forces are represented by circles. The fitted curves for the passive force are plotted ad dashed lines whereas the simplified model functions are represented by solid lines.

Defining $c_c = k_{c0}/k_{p0}c_{a,y}c_{f,y}c_{v,y}c_{n,y}$ and $c_f = k_{f0}/k_{p0}c_{a,x}c_{f,x}c_{v,x}c_{n,x}$, the desired relationships

$$\begin{aligned} k_f &= c_f k_p\,, \\ k_c &= c_c k_p \end{aligned}$$

for the specific forces are obtained, where k_p is given like before by

$$k_p = (k_{p0}c_a c_f)a_p^{-m_c} f^{-m_d} \left(\alpha_1 v_c^{\beta_1} + \alpha_2 v_c^{-\beta_2}\right) \left(\gamma_1 n^{\kappa_1} + \gamma_2 n^{-\kappa_2}\right).$$

Note that the index z is omitted because only this one force component is considered. Recalling (3.18), i.e.

$$F_p = k_p a_p f = \bar{F}_n\, a_p(t)^{1-m_c} \left(\frac{v_x(t)}{n}\right)^{1-m_d} \left(\alpha_1 v_c(t)^{\beta_1} + \alpha_2 v_c(t)^{-\beta_2}\right)$$

with $\bar{F}_n = k_{p0}c_{a,p}c_{f,p}G_n$, the corresponding deflection in z-direction is computed by

$$\begin{aligned} \delta &:= \delta_z = \frac{F_p}{k_{ez}}\,, \\ &= D\, a_p^{1-m_c} v_x^{1-m_d} \left(\alpha_1 v_c^{\beta_1} + \alpha_2 v_c^{-\beta_2}\right) \end{aligned} \tag{3.39}$$

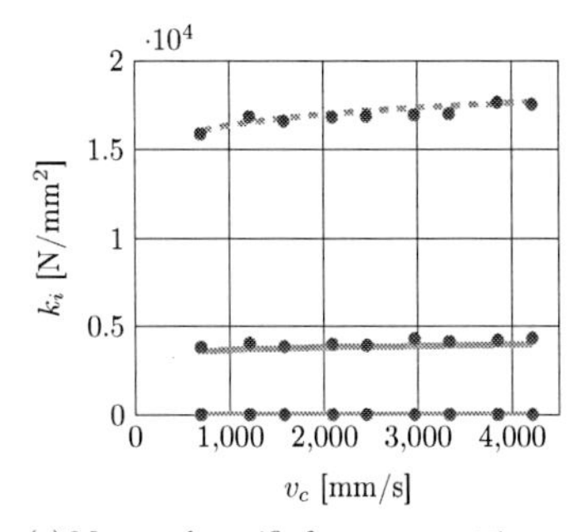

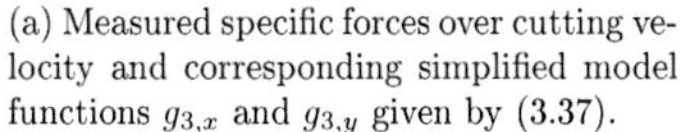

(a) Measured specific forces over cutting velocity and corresponding simplified model functions $g_{3,x}$ and $g_{3,y}$ given by (3.37).

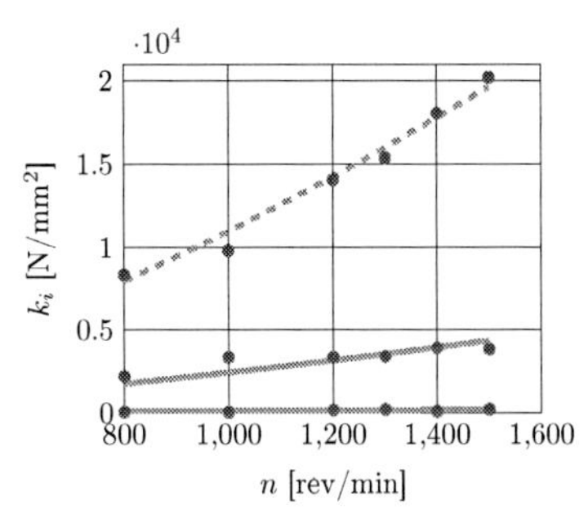

(b) Measured specific forces over rotational speed and corresponding simplified model functions $g_{4,x}$ and $g_{34,y}$ given by (3.38).

Figure 3.9: The measured specific forces are represented by circles, whereas the curves g_3 and g_4 for the passive force are plotted as dashed lines. In contrast the simplified model functions are plotted as solid lines.

with $D := \bar{F}_n k_{ez}^{-1} n^{m_d - 1}$. Therefore, the deflections in x- and $y-$direction are given by

$$\delta_x = \frac{F_f}{k_{ex}} = \frac{c_f F_p}{k_{ex}} = \left(c_f \frac{k_{ez}}{k_{ex}}\right) \delta =: C_x \delta \,, \tag{3.40}$$

$$\delta_y = \frac{F_c}{k_{ey}} = \frac{c_c F_p}{k_{ey}} = \left(c_c \frac{k_{ez}}{k_{ey}}\right) \delta =: C_y \delta \,. \tag{3.41}$$

The following procedure in order to derive the system of differential equations describing the process is the same as in Subsection 3.2.1.2 in the case for the complete force model. Plugging (3.39), (3.40) and (3.41) into the differential equations (3.22) - (3.24) yields

$$\dot{x}(t) = +v_f^0(t) - C_x \dot{\delta}(t) - \dot{\Delta}_x(t) \,, \tag{3.42}$$

$$\dot{y}(t) = - C_y \dot{\delta}(t) - \dot{\Delta}_y(t) \,, \tag{3.43}$$

$$\begin{aligned} \dot{z}(t) = & -\dot{a}_p^0(t) + \tfrac{1}{l_h}\left(C_x^2 + C_y^2\right)\delta(t)\dot{\delta}(t) + \dot{\delta}(t) - \dot{\Delta}_z(t) \\ & + \dot{x}(t)\tan\left(\beta_y(t)\right) + x(t)\frac{\dot{\beta}_y(t)}{\cos^2\left(\beta_y(t)\right)} \end{aligned} \tag{3.44}$$

for the derivatives of the positions. By (3.21) the derivative of the deflection is given by

$$\begin{aligned} \dot{\delta} = & D a_p(t)^{-m_c} v_x(t)^{-m_d} \left\{ \dot{v}_c(t) a_p(t) v_x(t) \left[\alpha_1 \beta_1 v_c(t)^{\beta_1 - 1} - \alpha_2 \beta_2 v_c(t)^{-\beta_2 - 1}\right] \right. \\ & \left. + \left[\alpha_1 v_c(t)^{\beta_1} + \alpha_2 v_c(t)^{-\beta_2}\right] \left[\left(1 - m_c\right)\dot{a}_p(t) v_x(t) + \left(1 - m_d\right)\dot{v}_x(t) a_p(t)\right] \right\} , \end{aligned}$$

and analogously to (3.26), an equation for $\dot{v}_x$ is derived to be

$$\begin{aligned} \dot{v}_x = & \frac{D^{-1}\dot{\delta} a_p^{m_c - 1} v_x^{m_d} + 2\pi n v_x^2 \left[\alpha_1 \beta_1 v_c^{\beta_1 - 1} - \alpha_2 \beta_2 v_c^{-\beta_2 - 1}\right]}{\left[\alpha_1 v_c^{\beta_1} + \alpha_2 v_c^{-\beta_2}\right]\left(1 - m_d\right)} \\ & - \frac{\left(1 - m_c\right)}{\left(1 - m_d\right)} \frac{\dot{a}_p}{a_p} v_x \,. \end{aligned} \tag{3.45}$$

Table 3.5: Model constants for simplified function $\tilde{g}_3$.

	Thrust force $i=t$	Cutting force $i=c$	Feed force $i=f$
$\tilde{g}_{3,i}(v_c)$	$\alpha=\alpha_t=0.65$	$\alpha_c=0.77$	$\alpha_f=0.05$
	$\beta=\beta_t=0.05$	$\beta_c=0.03$	$\beta_f=0.24$

Therefore, a new system of differential equations

$$\begin{aligned}\dot{v}_c &= -2\pi n v_x\,,\\ \dot{a}_p &= \dot{a}_p^0-\left(1+\frac{C_x^2+C_y^2}{l_h}\delta\right)\dot{\delta}+\dot{\Delta}_z-v_x\tan(\beta_y)+\frac{v_c}{2\pi n}\frac{\dot{\beta}_y}{\cos^2(\beta_y)}\,,\\ \dot{v}_x &= \frac{D^{-1}\dot{\delta}\,a_p^{m_c-1}v_x^{m_d}+2\pi n v_x^2\left[\alpha_1\beta_1 v_c^{\beta_1-1}-\alpha_2\beta_2 v_c^{-\beta_2-1}\right]}{\left[\alpha_1 v_c^{\beta_1}+\alpha_2 v_c^{-\beta_2}\right](1-m_d)}-\frac{(1-m_c)}{(1-m_d)}\frac{v_x}{a_p}\dot{a}_p\,,\\ \dot{\delta} &= C_x^{-1}\left(v_f^0-v_x-\dot{\Delta}_x\right)\end{aligned}$$

is derived. Introducing constants $a_1:=2\pi n$, $a_2:=\left(C_x^2+C_y^2\right)/l_h$, $a_3:=(DC_x)^{-1}$, $a_4:=1-m_d$, $a_5:=1-m_c$, $a_6:=a_5/a_4$, $a_7:=C_x^{-1}$, $a_8:=a_1\alpha_1\beta_1$, and $a_9:=a_1\alpha_2\beta_2$, this system can be formulated as

$$\dot{v}_c = -a_1 v_x\,, \tag{3.46}$$

$$\dot{a}_p = \dot{a}_p^0-(1+a_2\delta)\,\dot{\delta}+\dot{\Delta}_z-v_x\tan(\beta_y)+\frac{v_c\dot{\beta}_y}{a_1\cos^2(\beta_y)}\,, \tag{3.47}$$

$$\dot{v}_x = \frac{D^{-1}\dot{\delta}\,a_p^{-a_5}v_x^{1-a_4}+v_x^2\left[a_8 v_c^{\beta_1-1}-a_9 v_c^{-\beta_2-1}\right]}{a_4\left[\alpha_1 v_c^{\beta_1}+\alpha_2 v_c^{-\beta_2}\right]}-a_6\frac{v_x}{a_p}\dot{a}_p\,, \tag{3.48}$$

$$\dot{\delta} = a_7\left(v_f^0-v_x-\dot{\Delta}_x\right). \tag{3.49}$$

The next simplification is based on a modification of the function g_3. In Table 3.2 it can be observed that $\alpha_1 \gg \alpha_2$ and $\beta_1 \ll \beta_2$. Therefore, it is assumed that the first term in (3.4) is the dominant one and that g_3 can be approximated by

$$\tilde{g}_3(v_c) = \alpha v_c^{\beta}\,. \tag{3.50}$$

The corresponding values of the model constants determined again by a least-square fit to the measuremts are collected in Table 3.5. In Figure 3.10 the function $\tilde{g}_3$ is plotted over the cutting velocity and it is compared to measured forces as well as to the function g_3. As it can be seen, $\tilde{g}_3$ is a good approximation.

Using the simplified functions $\tilde{g}_3$ instead of g_3, the system of differential equations (3.46)-(3.49) can be simplified and rewritten as

$$\dot{v}_c(t) = -a_1 v_x(t)\,, \tag{3.51}$$

$$\begin{aligned}\dot{a}_p(t) = \dot{a}_p^0(t) &- a_7\left(1+a_2\delta(t)\right)\left(v_f^0(t)-v_x(t)-\dot{\Delta}_x(t)\right)+\dot{\Delta}_z(t)\\ &-v_x(t)\tan(\beta_y(t))+\frac{v_c(t)\dot{\beta}_y(t)}{a_1\cos^2(\beta_y(t))}\,,\end{aligned} \tag{3.52}$$

$$\dot{v}_x(t) = a_{10}\dot{\delta}(t)\,a_p(t)^{-a_5}v_x(t)^{1-a_4}v_c(t)^{-\beta}+a_{11}\frac{v_x(t)^2}{v_c(t)}-a_6\frac{v_x(t)}{a_p(t)}\dot{a}_p(t)\,, \tag{3.53}$$

$$\dot{\delta}(t) = a_7\left(v_f^0(t)-v_x(t)-\dot{\Delta}_x(t)\right), \tag{3.54}$$

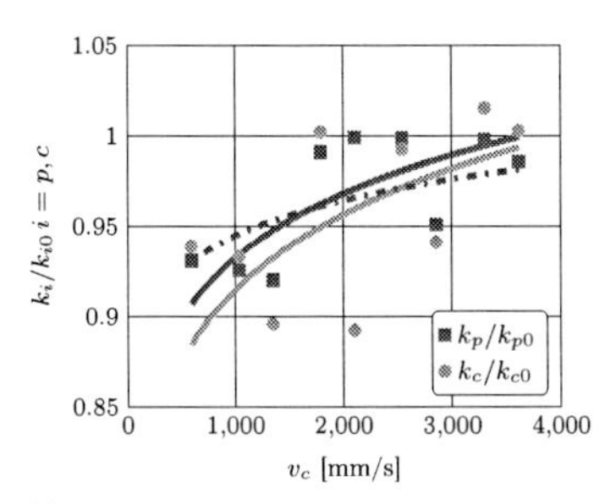

(a) Normalized specific forces over cutting velocity.

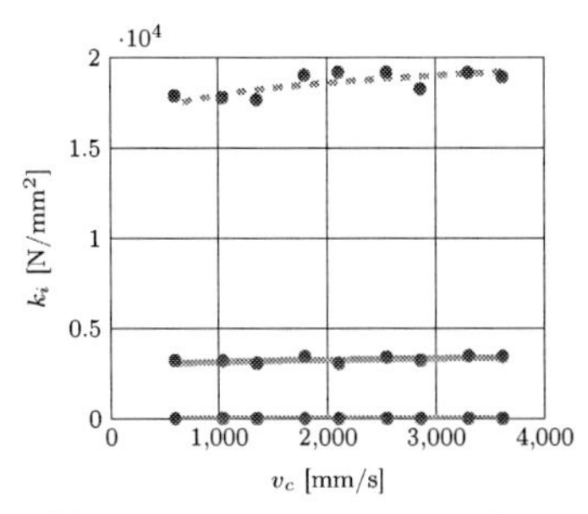

(b) Specific forces over cutting velocity.

Figure 3.10: Comparison between measured specific forces represented as circles and the model function $\tilde{g}_3$ given by (3.50). The curve is plotted as dashed line for the passive force and as a solid line for the other force components which are proportional to the first one.

where the abbreviations $a_{10} := 1/(Da_4\alpha)$ and $a_{11} := a_8/(a_4\alpha)$ are introduced.

Note that this system can be written as an explicitly given differential equation system

$$\dot{\mathbf{u}} = f(\mathbf{u}, \mathbf{p}^0, \mathbf{q}^0).$$

That means that the right hand side $f = f(\mathbf{u}, \mathbf{p}^0, \mathbf{q}^0)$ depends on the parameters of the process $\mathbf{p}^0 = (v_f^0, \dot{a}_p^0)$ and the machine $\mathbf{q}^0 = \left(\dot{\Delta}_x, \dot{\Delta}_y, \dot{\Delta}_z, \beta_y, \dot{\beta}_y\right)$ as well as on $\mathbf{u} = (v_c, a_p, v_x, \delta)$, but not on $\dot{\mathbf{u}}$. Hence, it can be rewritten as

$$\begin{aligned}
\dot{u}_1 =& -a_1 u_3 , \\
\dot{u}_2 =& p_1 - a_7 (1 + a_2 u_4) (p_2 - u_3 - q_1) + g(u_1, u_3, q_3, q_4) , \\
\dot{u}_3 =& a_{12} (p_2 - u_3 - q_1) u_1^{-\beta} u_2^{-a_5} u_3^{1-a_4} + a_{11} u_3^2 u_1^{-1} \\
& - a_6 \frac{u_3}{u_2} \left[p_1 - a_7 (1 + a_2 u_4) (p_2 - u_3 - q_1) + g(u_1, u_3, q_3, q_4, q_5)\right] , \\
\dot{u}_4 =& a_7 (p_2 - u_3 - q_1) ,
\end{aligned} \tag{ODE 2}$$

where the abbreviation $a_{12} := a_7 a_{10}$ and the function

$$g(u_1, u_3, q_3, q_4, q_5) := q_3 - u_3 \tan(q_4) + \frac{u_1 q_5}{a_1 \cos^2(q_4)}$$

are used. This system is called in the following simplified process model (ODE 2).

To solve this system of ordinary differential equations, initial conditions are necessary which can be deduced from the starting point of the derivations, i.e. from the equations (3.12)-(3.14). Using the introduced abbreviations a_i and the simplified force model, they can be rewritten as

$$x(t) = -r + \int_0^t v_f^0(s)\, \mathrm{d}s \quad -C_x\delta(t) - \Delta_x(t) , \tag{3.55}$$

$$y(t) = -C_y\delta(t) - \Delta_y(t) , \tag{3.56}$$

$$z(t) = -a_p^0(t) + \tfrac{a_2}{2}\delta^2(t) \quad +\delta(t) - \Delta_z(t) + x(t)\tan(\beta_y(t)) . \tag{3.57}$$

Using the formula for the cutting velocity

$$v_c(t) = -a_1 x(t)$$

and for the deflection

$$\delta(t) = D\, a_p^{a_5}(t)\, v_x^{a_4}(t)\, \alpha v_c^{\beta}(t)$$

yields at $t = 0$

$$\begin{aligned} x(0) &= -r - C_x\delta(0) - \Delta_x(0)\,, \\ a_p(0) &= a_p^0(0) - \frac{a_2}{2}\delta^2(0) - \delta(0) + \Delta_z(0) - x(0)\tan(\beta_y(0))\,, \\ v_c(0) &= -a_1 x(0)\,, \\ \delta(0) &= D\, a_p^{a_5}(0)\, \left(v_x^0(0)\right)^{a_4}\, \alpha v_c^{\beta}(0)\,. \end{aligned}$$

Choosing a value for $v_x(0) =: v_{x,0}$ and $a_p(0) =: a_{p,0} = a_p^0(0)$, two equations remain

$$\begin{aligned} v_c(0) &= -a_1\left(-r - C_x\delta(0) - \Delta_x(0)\right), \\ \delta(0) &= D\, a_{p,0}^{a_5}\, v_{x,0}^{a_4}\, \alpha v_c^{\beta}(0)\,, \end{aligned}$$

which can be solved numerically. Hence, as initial conditions for $\mathbf{u}_0 = \mathbf{u}(0)$ it has been derived that

$$\mathbf{u}_0 = \left(v_c(0),\, a_{p,0}, v_{x,0}, \delta(0)\right)^{\top}.$$

This simplified system of ordinary differential equations will be used in Chapter 4 for the analysis of the forward problem.

Summarizing the main results of this chapter, note that first a new micro-force model for ultra-precision turning has been developed. Using this force model in order to compute the actual process parameters by incorporating the influence of the vibrations of the workpiece and the elastic deflections of the tool holder caused by the process forces, the implicitly given system (ODE 1) of differential equations has been derived.

In the following, several simplifying assumption about the force model have been made, and each of these simplified force model results in a simplified process model. The result is thus a hierarchy of different force models with decreasing complexity, and corresponding process models, which are summarized in Table 3.6.

Remember that the differential equations are the consequence of introducing the actual feed velocity $v_x(t) = \dot{x}(t)$. In case of steady state conditions $v_x = v_f$ or when assuming that both velocities are approximately the same, the set of algebraic equations for the positions (3.12)-(3.14), for the actual depth of cut (3.15), and for the deflections (3.20) in all three directions would be sufficient in order to describe the process completely, i.e. no system of ordinary differential equations would thus be necessary.

The process model is coupled in Subsection 3.2.3 with the machine model introduced in the next subsection. It is also the basis for the parameter-to-state map and thus of the forward problem in Chapter 4.

Table 3.6: Hierarchy of the considered force models and the resulting ordinary differential equation systems.

Force model	Corresponding process model given by
1. Complete force model specific force for each component $i = c, p, f$ as in (3.6): $k_i = \bar{k}_{i,0}\, a_p^{-m_{c,i}} f^{-m_{d,i}} \left(\alpha_{1,i} v_c^{\beta_{1,i}} + \alpha_{2,i} v_c^{-\beta_{2,i}}\right)\left(\gamma_{1,i} n^{\kappa_{1,i}} + \gamma_{2,i} n_i^{-\kappa_{2,i}}\right)$	
1.1 for constant n during the process, compare with (3.18): $k_i = \bar{k}_{i,0} G_{n,i}\, a_p^{-m_{c,i}} f^{-m_{d,i}} \left(\alpha_{1,i} v_c^{\beta_{1,i}} + \alpha_{2,i} v_c^{-\beta_{2,i}}\right)$	(ODE 1), see page 62
2. Assume $g_2(f)$ to be constant, i.e. $m_{d,i} = 0$ for $i = c, p, f$: $k_i = \bar{k}_{i0} G_{n,,i}\, a_p^{-m_{c,i}} \left(\alpha_{1,i} v_c^{\beta_{1,i}} + \alpha_{2,i} v_c^{-\beta_{2,i}}\right)$	(ODE 1) with $m_{d,i} = 0$
3. Assume proportional relationship: $k_c = c_c k_p$ and $k_f = c_f k_p$ (see (3.35)-(3.38)) with specific passive force given for constant n by $k_p = \bar{k}_{p,0} G_n\, a_p^{-m_c} f^{-m_d} \left(\alpha_1 v_c^{\beta_1} + \alpha_2 v_c^{-\beta_2}\right)$	(3.46)-(3.49), see page 68
4. Simplified model: Simplify function $g_3(v_c)$ by assuming as in (3.50) that $\tilde{g}_3(v_c) = \alpha v_c^{\beta}$, i.e. the specific passive force is given for constant n by $k_p = \left(\bar{k}_{p,0} G_n \alpha\right) a_p^{-m_c} f^{-m_d} v_c^{\beta}, \qquad k_c = c_c k_p$, and $k_f = c_f k_p$	(ODE 2), see page 69

3.2.2 The machine model

In this section a machine model is derived which can be used to predict machine vibrations of the test stand caused by unbalance distributions on the main spindle. Since the experimental platform from Figure 3.2 has a complex structure, which may be difficult to model, simplifying assumptions have to be taken in order to be able to handle the model mathematically. Nevertheless, the simplified model has still to be a good approximation of the reality. The resulting model has been published several times, see for example the references [10, 42, 43], and the remainder of this subsection is a summary of [10].

3.2.2.1 Derivation of the vibration equation for the machine

In order to develop a model for the machine structure, the platform is first divided into several components, which are illustrated in Figure 3.11. These components are the rotating part of the spindle, the spindle casing, the rotating part of the engine, and the engine casing. The spindle rotor and casing are connected by an air bearing with two spherical calottes which are modeled by two spring-damper elements, although in a first attempt any damping in the springs is neglected. The engine bearings are also modeled as spring-damper elements. Spindle and engine are connected by a coupling that can compensate deformations in axial and radial directions and is also modeled as a spring element. Spindle and engine are supported by a granite board that is assumed to be rigid. The joints to the granite board are modeled as firm spring elements. The coordinate system for the machine model is the same as for the process model, see Figure 3.11. The spindle rotates counterclockwise around the z- axis.

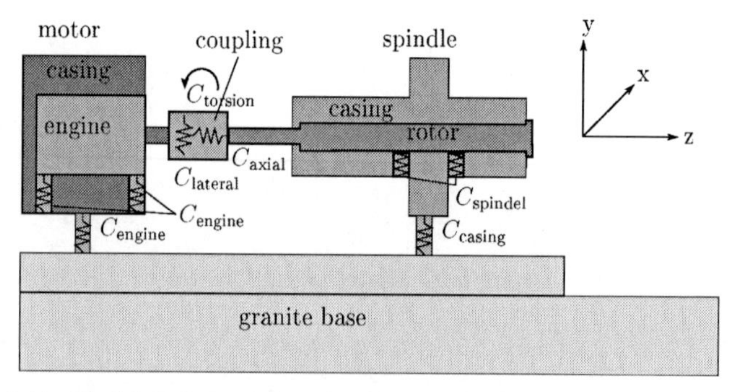

Figure 3.11: Modeled parts of the experimental platform: spindle, spindle casing, rotor, and engine casing. Each part is divided into several elements, see [42].

Secondly, a vibration (or displacement) model for each part of the machine is developed separately. If unbalances would be the only cause for vibrations, it would suffice to allow vibrations only in radial direction because there would be no vibrations in z- direction. Since unbalances excite harmonic vibrations, the vibrations in

x and y direction are the same except for a phase shift of $\pi/2$. Nevertheless, the forces from the cutting process act in all three directions and therefore can excite vibrations in all three directions. Thus, each point along the z-axis, i.e. each beam element of infinitesimal length ∂z, has the following degrees of freedom (DOF): the displacement u, v, w in x, y, z-direction, the torsion angle β_z, and the cross section slopes β_x and β_y, see Figure 3.3 for an illustration. The DOF are collected in a vector $\boldsymbol{\lambda} = \boldsymbol{\lambda}(x,t)$. The computation of $\boldsymbol{\lambda}$ is based on an energy formulation, the so called Principle of Virtual Displacements, see [23], Chapter 5. This principle is equivalent to equilibrium conditions from which a partial differential equation for $\boldsymbol{\lambda}(x,t)$ can be derived. Starting from the energy formulation, the Finite Element Method (FEM) is used for the discretization in the space variable z. The result is a system of ordinary differential equations in time of the form

$$\mathbf{M}\ddot{\boldsymbol{\lambda}}(t) + \mathbf{S}\boldsymbol{\lambda}(t) = \mathbf{p}_{unb}(t), \tag{3.58}$$

where $\mathbf{M}$ denotes the mass matrix, $\mathbf{S}$ the stiffness matrix, and $\mathbf{p}_{unb}$ a load vector caused by unbalance distributions. In case of damping in the system, a third term $\mathbf{D}\dot{\boldsymbol{\lambda}}(t)$ with a sparse damping matrix $\mathbf{D}$ has to be included on the left hand side of the equation.

For the discretization, the considered platform parts are divided into finite beam elements with nodes at each end. The movement of each point between the nodes is described by ansatz functions scaled with the movement of the nodes. Consideration of the boundary and transition conditions between the end node of one element and the first node of the next element provides system matrices $\mathbf{M}$ and $\mathbf{S}$ for each part of the platform. This procedure is well known, see [23], and explained in detail in [31] for the case of a model for a micro-milling machine. The design of the spindle, the casing, and the engine suggested the following numbers of elements:

- Spindle rotor with the coupling: 36 elements,
- Spindle casing: 30 elements,
- Rotational part of the engine: 3 elements,
- Engines casing: 2 elements.

As introduced above, each node has 6 DOF: the displacements u, v, w in x, y, z-direction, the torsion angle β_z, and the cross section slopes β_x and β_y. The DOF of each of the parts specified above are collected in vectors $\boldsymbol{\lambda}_{sp-rot}$, $\boldsymbol{\lambda}_{sp-cas}$, $\boldsymbol{\lambda}_{e-rot}$ and $\boldsymbol{\lambda}_{e-cas}$. A discretization into N_i elements in our model leads to $N_i + 1$ nodes and thus $6 \cdot (N_i + 1)$ DOF in the model of the ith part. The vectors of the DOF are subject to Equation (3.58) with mass matrices $\mathbf{M}_{sp-rot}$, $\mathbf{M}_{sp-cas}$, $\mathbf{M}_{e-rot}$, $\mathbf{M}_{e-cas}$ and stiffness matrices $\mathbf{S}_{sp-rot}$, $\mathbf{S}_{sp-cas}$, $\mathbf{S}_{e-rot}$ and $\mathbf{S}_{e-cas}$. If we collect all DOF in one vector

$$\boldsymbol{\lambda} = (\boldsymbol{\lambda}_{sp-rot}^T, \boldsymbol{\lambda}_{sp-cas}^T, \boldsymbol{\lambda}_{e-rot}^T, \boldsymbol{\lambda}_{e-cas}^T)^T,$$

we get a block diagonal structure for the entire mass matrix

$$\mathbf{M} = \begin{pmatrix} \mathbf{M}_{sp-rot} & 0 & 0 & 0 \\ 0 & \mathbf{M}_{sp-cas} & 0 & 0 \\ 0 & 0 & \mathbf{M}_{e-rot} & 0 \\ 0 & 0 & 0 & \mathbf{M}_{e-cas} \end{pmatrix}.$$

The stiffness matrix $\mathbf{S}$ is arranged in the same way. So far, the matrices $\mathbf{M}$ and $\mathbf{S}$ have a block diagonal structure and the dimension 450×450. In the stiffness matrix, entries, corresponding to the bearing elements that relate the DOF of the nodes in the connected parts via a stiffness parameter C and damping parameters D, have to be added. This results in off-diagonal elements for the entire stiffness matrix and in the damping matrix $\mathbf{D}$, respectively.

At last the machine model is adjusted, and unknown model parameters are identified by comparing modeled and measured vibration measurements at two sensor positions on the spindle casing. For more information about the model calibration, see [10, 43].

3.2.2.2 Solution of the vibration equation in the presence of unbalances

If only unbalances in the turning lathe are considered, the right hand side $\mathbf{p}_{unb}(t)$ of (3.58) has harmonic entries depending on the angular velocity $\omega = 2\pi n$. An unbalance is modeled as a mass Δm displaced from the shaft by a vector $\mathbf{r}_m = r_m e^{i\cdot\phi}$. Here ϕ is the angle to a given zero mark. If the displaced mass rotates with angular velocity ω, it induces a harmonic centrifugal force (or load)

$$\mathbf{p}_{unb}(t) = \omega^2 \bar{\mathbf{p}}_0 e^{i\omega t} \quad \text{with } \bar{\mathbf{p}}_0 := \Delta m r_m e^{i\cdot\phi}.$$

It is assumed that the induced vibration is also harmonic with the same frequency ω, i.e. $\boldsymbol{\lambda}(t) = \boldsymbol{\lambda}_0 e^{i\omega t}$ with a complex and time independent amplitude $\boldsymbol{\lambda}_0$. Inserting this in (3.58) provides a time independent expression

$$\boldsymbol{\lambda}_0 = (-\omega^2 \mathbf{M} + \mathbf{S})^{-1} \bar{\mathbf{p}}_0 \tag{3.59}$$

for the vibration amplitude of the vibration $\boldsymbol{\lambda}_0$. Therefore, $\boldsymbol{\lambda}$ is determined via $\boldsymbol{\lambda}_0$.

However, in the presence of cutting forces that act independently in all three directions x, y and z, the complex notation is not preferable anymore. Therefore, the complex force $\mathbf{p}_{unb}(t)$ is projected onto the x- and y- axis as follows

$$\begin{aligned} F_x &= \omega^2 \Delta m r_m \sin(\omega t + \phi) = \mathrm{Im}(\mathbf{p}_{unb}(t)), \\ F_y &= \omega^2 \Delta m r_m \cos(\omega t + \phi) = \mathrm{Re}(\mathbf{p}_{unb}(t)). \end{aligned}$$

Those forces only apply to the displacement DOF in x- and y-direction. All the other DOF are not affected. Therefore, the sub-vector $\mathbf{p}_k$ of $\mathbf{p}_{unb}$ containing the entries for the DOF of the k-th node, $k = 1, \cdots, N$, has the form

$$\mathbf{p}_k = \begin{pmatrix} \mathrm{Im}(\omega^2 \Delta m_k r_{m,k} e^{i\phi_k} e^{i\omega t}) \\ \mathrm{Re}(\omega^2 \Delta m_k r_{m,k} e^{i\phi_k} e^{i\omega t}) \\ 0 \\ 0 \\ 0 \\ 0 \end{pmatrix}.$$

The vector $\mathbf{p}_{unb}(t) = (\mathbf{p}_k)_k$ is split into parts with sin and cos entries, i.e.

$$\mathbf{p}_{unb} = \mathrm{Im}(\mathbf{q}^1 e^{i\omega t}) + \mathrm{Re}(\mathbf{q}^2 e^{i\omega t}), \quad \text{with } \mathbf{q}^{1,2} = (\mathbf{q}_k^{1,2})_k$$

and

$$\mathbf{q}_k^1 = \begin{pmatrix} \mathrm{Im}(\omega^2 \Delta m_k r_{m,k} e^{i\phi_k} e^{i\omega t}) \\ 0 \\ 0 \\ 0 \\ 0 \\ 0 \end{pmatrix}, \quad \mathbf{q}_k^2 = \begin{pmatrix} 0 \\ \mathrm{Re}(\omega^2 \Delta m_k r_{m,k} e^{i\phi_k} e^{i\omega t}) \\ 0 \\ 0 \\ 0 \\ 0 \end{pmatrix},$$

such that the right hand side ansatz can be applied to each of these parts in order to solve (3.58). Inserting the equation $\boldsymbol{\lambda}_{unb}^j(t) = \boldsymbol{\lambda}^j e^{i\omega t}$, $(j = 1, 2)$ and its second derivative in (3.58) yields

$$\begin{aligned} \boldsymbol{\lambda}_{unb}(t) &= \boldsymbol{\lambda}_{unb}^1(t) + \boldsymbol{\lambda}_{unb}^2(t), \\ &= \mathrm{Im}((-\omega^2 \mathbf{M} + \mathbf{S})^{-1} \mathbf{q}_1 e^{i\omega t}) + \mathrm{Re}((-\omega^2 \mathbf{M} + \mathbf{S})^{-1} \mathbf{q}_2 e^{i\omega t}). \quad (3.60) \end{aligned}$$

The solution of (3.58) is the sum of the particular solution $\boldsymbol{\lambda}_{unb}$ and the general solution of the homogeneous equation. The latter will be given in Subsection 3.2.3. After a certain time of rotation with a constant angular velocity and no other forces than those from unbalances, the homogeneous solution will die out due to small damping effects. Hence, in this case the entire solution is $\boldsymbol{\lambda}_{unb}$.

To summarize, the machine model can be used to calculate the vibrations $\boldsymbol{\lambda}$ caused by an unbalance distribution given as a mass at a certain radial distance and angle position by solving the ordinary differential equation (3.58). In a more abstract way, the presented model can be regarded as a forward model with linear operator A mapping an unbalance to the resulting vibrations. Solving the corresponding inverse problem, or in other words inverting the operator A by use of the regularization techniques presented in Subsection 2.4, offers the possibility to compute balancing weights and positions for the compensation of the unbalance. This approach will not be presented in this thesis, but has been published in [10, 11] and [3].

3.2.3 The coupled interaction model

In the last two sections, two sub-models have been developed, one computing the forces of the cutting process and displacements of the tool holder, and one describing the vibration behavior of the cutting machine. Both models have been built separately, but in practice they influence each other. Developing the process model, the influence of the displacements of the workpiece which is the first node in the machine model has already been considered, see the displacements Δ_i, $(i = x, y, z)$ and the angle β_y in the equations (3.12)-(3.14) for the tool position. However, the cutting forces themselves cause a force and a moment acting on the workpiece which lead to an additional load in the vibration equation (3.58). Therefore, both sub-models have to be coupled which is done in detail in this subsection.

Starting point is the additional load vector from the cutting forces and moments which has to be added in the vibration equation (3.58). Recall that a moment is defined as the cross product of the lever arm, connecting the rotating axis to the point of force application, and the force. In our case that is

$$M_i = r_a \times F_i, \quad i = (p, c) \quad \text{and } M_f = 0$$

with length r_a of the lever arm given by the tool position as $r_a := \sqrt{x^2+y^2}$. The forces F_i are directly computed by multiplication of the displacements δ_i, which are part of the solution vector $\mathbf{u}$ of the process model, with the corresponding stiffness k_{ei}. Considering the direction of each force component and the principle that the force acting on the tool is the negative force acting on the workpiece, the first six entries of the additional load vector are given by

$$\mathbf{p}_{cut}(t,\boldsymbol{\lambda},\mathbf{u}) = \left(F_f\left(\mathbf{u}\right), F_c\left(\mathbf{u}\right), -F_p(\mathbf{u}), 0, M_t\left(\mathbf{u}\right), -M_c\left(\mathbf{u}\right)\right)^T, \tag{3.61}$$

whereas the remaining entries are all zero.

Since the displacements of the workpiece equal the first three entries of $\boldsymbol{\lambda}$ and the rotations the next three entries, it holds that

$$(\Delta_x, \Delta_y, \Delta_z, \beta_x, \beta_y, \beta_z) = (\lambda_1, \lambda_2, \lambda_3, \lambda_4, \lambda_5, \lambda_6).$$

Moreover, the right-hand side f of the process model (3.33) includes these displacements collected in the parameter $\mathbf{q}^0$. The solution of the process model depends thus on $\boldsymbol{\lambda}$ and therefore also on the forces and the load vector $\mathbf{p}_{cut}$. The right hand side of the vibration equation (3.58) depends thus also on its solution, and we have to deal with a nonlinear problem which cannot be solved explicitly. Summarizing, the process machine interaction is described by a coupled system of ordinary differential equations

$$\begin{aligned} \mathbf{M}\ddot{\boldsymbol{\lambda}}(t) + \mathbf{S}\boldsymbol{\lambda}(t) &= \mathbf{p}_{unb}(t) + \mathbf{p}_{cut}(t, \boldsymbol{\lambda}(t), \mathbf{u}(t)), && (3.62)\\ f(\mathbf{u}(t), \dot{\mathbf{u}}(t), \mathbf{p}^0(t), \mathbf{q}^0(\boldsymbol{\lambda}(t))) &= 0 && (3.63) \end{aligned}$$

with parameter vector $\mathbf{p}^0(t) = \left(\dot{a}_p^0(t), v_f^0(t)\right)$ and unbalance distribution $\mathbf{p}_{unb}(t)$.

In order to solve the system approximately, a time step algorithm is employed. We assume the forces and moments from the cutting process to be constant during a small time interval, i.e. $\mathbf{p}_{cut}(t) = \mathbf{p}_{cut}(t_i)$ for $t \in [t_i, t_i + \Delta t]$. The vibrations of the machine are then computed by

$$\boldsymbol{\lambda}(t) = \mathbf{A}(\mathbf{p}_{unb}(t) + \mathbf{p}_{cut}(t_i)), \quad t \in [t_i, t_i + \Delta t],$$

where $\mathbf{A}$ describes the solution operator of (3.62). The resulting displacement and angles λ_i $(i = 1, 2, \ldots 6)$ are used to compute $\mathbf{q}^0$ via (3.34), and then $\mathbf{q}^0$ is plugged into the process model (3.63). After solving the process model in the interval $[t_i, t_i+\Delta t]$ assuming now constant machine displacements, the forces are determined from the solution via (3.19), i.e.

$$(F_f(t_{i+1}), F_c(t_{i+1}), F_p(t_{i+1})) = \mathbf{B}(\boldsymbol{\lambda}(t_i + \Delta t)),$$

where $\mathbf{B}$ is the combination of the solution operator of the process model and the operator which maps the solution to the forces via (3.19). At last, the load vector $\mathbf{p}_{cut}$ is computed by (3.61), and the procedure restarts by solving the vibration equation for $t \in [t_{i+1}, t_{i+1} + \Delta t]$. This routine is repeated until the end of the desired time interval $t \in [t_0, t_{end}]$, for which the combined system should be solved, is reached.

3.2.3.1 Solution of the vibration equation

Whereas the ordinary differential equation of the process model is solved numerically, the vibration equation is solved directly what means like solving by hand. In the following, this solution method taken from [10] is presented. The entire solution is the sum

$$\boldsymbol{\lambda}(t_j) = \boldsymbol{\lambda}_h(t_j) + \boldsymbol{\lambda}_{cut}(t_j) + \boldsymbol{\lambda}_{unb}(t_j)$$

of the homogenous solution $\boldsymbol{\lambda}_h$ and the particular solutions $\boldsymbol{\lambda}_{cut}$ and $\boldsymbol{\lambda}_{unb}$ for the right hand sides $\mathbf{p}_{cut}$ and $\mathbf{p}_{unb}$. How to determine the latter solution has already shown in Section 3.2.2. In order to compute the remaining two solution parts, first the Eigenvalue decomposition

$$\mathbf{M}^{-1}\mathbf{S} = \mathbf{V}\boldsymbol{\Phi}\mathbf{V}^{-1}$$

of $\mathbf{M}^{-1}\mathbf{S}$ is computed, which can be done with MATLAB. Here $\boldsymbol{\Phi}$ denotes the diagonal matrix of eigenvalues μ_j and $\mathbf{V}$ the corresponding matrix of eigenvectors as columns. Multiplying the vibration equation from the left with $\mathbf{V}^{-1}\mathbf{M}^{-1}$ and using the eigenvalue decomposition, it follows that

$$\mathbf{V}^{-1}\ddot{\mathbf{u}} + \boldsymbol{\Phi}\mathbf{V}^{-1}\mathbf{u} = \mathbf{V}^{-1}\mathbf{M}^{-1}\mathbf{p}_{cut}.$$

Introducing $\mathbf{q} := \mathbf{V}^{-1}\mathbf{u}$ and $\mathbf{c} := \mathbf{V}^{-1}\mathbf{M}^{-1}\mathbf{p}_{cut}$, the decoupled system

$$\ddot{\mathbf{q}} + \boldsymbol{\Phi}\mathbf{q} = \mathbf{c}$$

is obtained, which can easily solved for constant $\mathbf{c}$. If follows for $\boldsymbol{\lambda}_{cut}$ that

$$q_j(t) = \frac{c_j}{\mu_j}\left(1 - \cos\left(\sqrt{\mu_j}\cdot t\right)\right) \quad , 1 \leq j \leq n$$

and hence that

$$\boldsymbol{\lambda}_{cut}(t) = \mathbf{V}\mathbf{q}(t) = \mathbf{V}[\mathbf{I} - \cos(\boldsymbol{\Phi}^{1/2}t)]\boldsymbol{\Phi}^{-1}\mathbf{c} \tag{3.64}$$

with diagonal matrices

$$\cos(\boldsymbol{\Phi}^{1/2}t) := \left(\cos\left(\sqrt{\mu_j}\,t\right)\right)_{j=1}^{N} \text{ and } \boldsymbol{\Phi}^{-1} := (1/\mu_j)_{j=1}^{N}$$

and $\sin(\boldsymbol{\Phi}^{1/2}t)$ analogously defined. The homogeneous solution $\boldsymbol{\lambda}_h$ is the real part of the complex solution

$$\boldsymbol{\lambda}_h^c(t) = \mathbf{V}\left[\exp(i\sqrt{\boldsymbol{\Phi}}t)\mathbf{A}^+ + \exp(-i\sqrt{\boldsymbol{\Phi}}t)\mathbf{A}^-\right]$$

with

$$\mathbf{A}^{\pm} = \begin{pmatrix} A_1^{\pm} \\ \vdots \\ A_N^{\pm} \end{pmatrix}, \quad \exp(\pm i\sqrt{\boldsymbol{\Phi}}t) := \begin{pmatrix} \exp(\pm i\sqrt{\lambda_1}t) & \cdots & 0 \\ \vdots & \ddots & \vdots \\ 0 & \cdots & \exp(\pm i\sqrt{\lambda_N}t) \end{pmatrix}.$$

The coefficients $\mathbf{A}^{\pm}$ have to be computed using the initial values $\boldsymbol{\lambda}_h^c(t_0)$ and the derivative

$$\dot{\boldsymbol{\lambda}}_h^c(t_0) = \mathbf{V}\left[i\exp(i\sqrt{\boldsymbol{\Phi}}t_0)\sqrt{\boldsymbol{\Phi}}\mathbf{A}^+ - i\exp(-i\sqrt{\boldsymbol{\Phi}}t_0)\sqrt{\boldsymbol{\Phi}}\mathbf{A}^-\right] \tag{3.65}$$

of $\boldsymbol{\lambda}_h^c(t_0)$. Therefore, the system

$$\begin{pmatrix} \mathbf{u}_h^c(t_0) \\ \dot{\mathbf{u}}_h^c(t_0) \end{pmatrix} = \begin{pmatrix} \mathbf{V}\exp(i\sqrt{\boldsymbol{\Phi}}t_0) & \mathbf{V}\exp(-i\sqrt{\boldsymbol{\Phi}}t_0) \\ i\mathbf{V}\exp(i\sqrt{\boldsymbol{\Phi}}t_0)\sqrt{\boldsymbol{\Phi}} & -i\mathbf{V}\exp(-i\sqrt{\boldsymbol{\Phi}}t_0)\sqrt{\boldsymbol{\Phi}} \end{pmatrix}\begin{pmatrix} \mathbf{A}^+ \\ \mathbf{A}^- \end{pmatrix} \tag{3.66}$$

for $\mathbf{A}^+$ and $\mathbf{A}^-$ has to be solved.

3.2.3.2 The solution algorithm

As explained above, the coupled interaction model is solved via a time step algorithm, i.e. for each time step the sub-models are solved separately. Assuming that the spindle is rotating with a constant angular frequency ω, an unbalance distribution is set which results in a load $\mathbf{p}_{unb}$. Furthermore, it is assumed that the homogeneous solution is died out before the cutting process starts at $t = t_0$. Consequently, at $t = t_0$ the forces of the cutting process are zero, and the initial solution of the vibration equation and its derivative are given by

$$\begin{aligned} \boldsymbol{\lambda}(t_0) &= \boldsymbol{\lambda}_{unb}(t_0) = \Im((-\omega^2\mathbf{M}+\mathbf{S})^{-1}\mathbf{q}_1 e^{i\omega t_0}) + \Re((-\omega^2\mathbf{M}+\mathbf{S})^{-1}\mathbf{q}_2 e^{i\omega t_0}), \\ \dot{\boldsymbol{\lambda}}(t_0) &= \dot{\boldsymbol{\lambda}}_{unb}(t_0) = \Im(i\omega(-\omega^2\mathbf{M}+\mathbf{S})^{-1}e^{i\omega t_0})) + \Re(i\omega(-\omega^2\mathbf{M}+\mathbf{S})^{-1}e^{i\omega t_0}), \end{aligned}$$

compare with (3.60). Then the cutting process starts, and for a small time interval $[t_0, t_1]$ the vibrations of the machine are assumed to be constant. For given input process parameters $\mathbf{p}^0 = \left(\dot{a}_p^0, v_f^0\right)$ and computed $\mathbf{q}^0 = \left(\dot{\lambda}_1, \dot{\lambda}_2, \dot{\lambda}_3, \lambda_5, \dot{\lambda}_5\right)$, the ODE (3.63) of the process model is solved numerically in the interval $[t_0, t_1]$. Then the resulting load $\mathbf{p}_{cut}(t_1)$ is computed and is assumed to be constant during the next time interval $[t_0, t_1]$. The corresponding solution $\boldsymbol{\lambda}_{cut}$ is determined by (3.64). In order to compute the homogeneous solution $\boldsymbol{\lambda}_h$ via (3.65), the system for the initial values (3.66), i.e.

$$\begin{pmatrix} \mathbf{A}^+ \\ \mathbf{A}^- \end{pmatrix} = \begin{pmatrix} e^{i\boldsymbol{\Phi}^{1/2}t_0} & + \; e^{-i\boldsymbol{\Phi}^{1/2}t_0} \\ i\,e^{i\boldsymbol{\Phi}^{1/2}t_0}\boldsymbol{\Phi}^{1/2} & - \; i\,e^{-i\boldsymbol{\Phi}^{1/2}t_0}\boldsymbol{\Phi}^{1/2} \end{pmatrix}^{-1} \begin{pmatrix} \boldsymbol{\lambda}_h(t_0) \\ \dot{\boldsymbol{\lambda}}_h(t_0) \end{pmatrix}$$

must be solved for $\boldsymbol{\lambda}_h(t_0) = \boldsymbol{\lambda}(t_0) - \boldsymbol{\lambda}_{cut}(t_0) - \boldsymbol{\lambda}_{unb}(t_0)$. This results in the solution

$$\boldsymbol{\lambda}(t) = \boldsymbol{\lambda}_h(t) + \boldsymbol{\lambda}_{unb}(t) + \boldsymbol{\lambda}_{cut}(t), \quad t \in [t_0, t_1].$$

The procedure restarts with the computation of $\mathbf{p}_{cut}(t_2)$ by solving the process model in the interval $[t_1, t_2]$. To summarize to complete algorithm, a pseudo code is shown in Algorithmus 3.1.

3.2.3.3 Numerical example

The algorithm has been tested for several parameter settings. As expected, the presence of unbalances mainly affects the vibration amplitudes of the workpiece in radial direction x and y. Nevertheless, the deflection in z direction is affected, too. Quantitative effects can be observed for unbalance distributions of different magnitude. Here, only two examples are presented with the parameters and unbalance distributions summarized in Table 3.7. Note that the first entry in the vector $\mathbf{p}_0$ belongs to an unbalance load at the workpiece whereas the second and the third one are related with the balancer planes. Therefore, $\mathbf{p}_0$ is a sub-vecor of $\bar{\mathbf{p}}_0$ containing the load entries which correspond to the nodes where unbalances can be set. The setting 2 corresponds to a screw of 7.023g fixed at the workpiece holder at a radius of 87.5 mm.

The vibrations of the workpiece over time for both unbalance settings are shown in Figure 3.12. As expected, the screw at the workpiece holder induces radial vibrations of the workpiece. But even in the ideal case without any unbalance, there are

Algorithmus 3.1 Pseudo code of the process machine interaction model.

Require: Unbalance $\mathbf{p}_{unb}$, process parameter $\mathbf{p}^0 = \left(\dot{a}_p^0, v_f^0\right)$
  Initialization for t_0
    $\boldsymbol{\lambda}_0 := \boldsymbol{\lambda}(t_0) = \boldsymbol{\lambda}_{unb}(t_0)$ by (3.60)
    $\mathbf{q}^0 := \left(\dot{\lambda}_{0,1}, \dot{\lambda}_{0,2}, \dot{\lambda}_{0,3}, \lambda_{0,5}, \dot{\lambda}_{0,5}\right)$
  repeat
    $t_{i+1} := t_i + \Delta t$
    Compute

$$\mathbf{F}_{i+1} := (F_f(t_{i+1}), F_c(t_{i+1}), F_p(t_{i+1})) = B(\boldsymbol{\lambda}_0(t_i + \Delta t))$$

    and $\mathbf{p}_{cut}(t_{i+1})$
      by solving for $t \in [t_i, t_{i+1}]$

$$f(\mathbf{u}(t), \dot{\mathbf{u}}(t), \mathbf{p}^0(t), \mathbf{q}^0\,(t, \boldsymbol{\lambda})) = 0$$

      and using (3.19)
    Save $\mathbf{u}_i(t_{i+1})$ {as initial condition for the next interval}
    Compute

$$\boldsymbol{\lambda}_{i+1} := \boldsymbol{\lambda}\,(t_{i+1}) = \boldsymbol{\lambda}_{unb}\,(t_{i+1}) + \boldsymbol{\lambda}_{cut}\,(t_{i+1}) + \boldsymbol{\lambda}_h\,(t_{i+1})$$

    by
      $\boldsymbol{\lambda}_{unb}\,(t_{i+1})$ via (3.60)
      $\boldsymbol{\lambda}_{cut}\,(t_{i+1})$ via (3.64)
      $\boldsymbol{\lambda}_h\,(t_{i+1})$via (3.65) and (3.66)
    $\mathbf{q}^0 := \left(\dot{\lambda}_{i+1,1}, \dot{\lambda}_{i+1,2}, \dot{\lambda}_{i+1,3}, \lambda_{i+1,5}, \dot{\lambda}_{i+1,5}\right)$
  until $t_{i+1} = t_{end}$

Table 3.7: Settings for diamond turning simulations.

parameter		setting 1	setting 2	setting 3
unbalance	$\mathbf{p}_0$	$[0, 0, 0]$	$[61.45, 0, 0]$	$[159, 0, 0]$
rational speed	n	1500 rev/min	1500 rev/min	1000 rev/min
feed rate	f	12 μm/min	12 mm/rev	8.33 mm/rev
depth of cut	a_p	5 μm	5 μm	5 μm
radius	r	30 mm	30 mm	30 mm
time step	Δt	1 ms	1ms	1ms

very small vibrations less than one picometer observable in radial as well as axial direction, which are caused by the process forces.

The corresponding forces and the resulting depth of cut are also affected by the vibrations of the workpiece like shown in Figure 3.13. The amplitudes of the vibrations and hence their influence on the forces and depth of cut depend on the stiffness of the machine and the set mass load of the unbalances. Since the machine model do not contain any damping, the vibrations are not damped in the simulations. In order to be able to compare the simulations with measurements, first damping effects have to be included in the machine model and secondly the correct stiffness values have to be determined experimentally. More examples for different unbalance distributions are shown in [9, 7, 10].

3.3 Simulation of the machined surface

The visualization of the machined surfaces is important for the investigation of the influence of unbalances on the surface quality because surface generation in diamond turning is affected by relative tool positioning errors. After having solved the machine process interaction model, the computed vibrations of the workpiece and the tool are used for the simulation of the resulting surface. The surface simulation provides the global form as well as the surface micro-topography. It is consequently possible to compute surface characterizing parameters like form deviation or roughness. In this subsection, an approach for the surface simulation as well as computed examples are presented. It is based on a surface generation model for micro cutting processes with geometrically defined cutting edges [46, 60].

Figure 3.14 shows a scheme of the surface simulation. In a first step, the interaction model is solved numerically like explained in the previous subsection. The solution contains the actual process parameters and deflections of the tool as well as the vibrations and rotations of the workpiece. With help of this information, the position model allows the computation of the whole tool trajectory or in other words the computation of the tool position relative to the workpiece. The position is described by affine linear transformations, namely translation and rotation, expressed as homogeneous matrices. Since only the relative position of tool and workpiece is relevant for the material removal model, the workpiece motions are transferred to the tool. Consequently, output of the position model is the relative tool position operator Φ.

In the last step, the material of the workpiece which is removed by the tool is determined. The material removal is computed under idealized conditions, assuming an isotropic homogeneous material and an ideal sharp cutting edge. In other words, the material which passed by the tool edge is removed completely, and ploughing as well as elastic recovery effects are neglected. Also material induced vibrations, see [39], are not considered in the surface generation.

In Figure 3.15 some example of computed surfaces and roughness profiles are shown. The simulated situation corresponds to setting 3 in Table 3.1 at the bottom and the same parameter setting but without any unbalance on the top of the figure. A more detailed description of the surface generation algorithm and interpretation of the simulations can be found in [7].

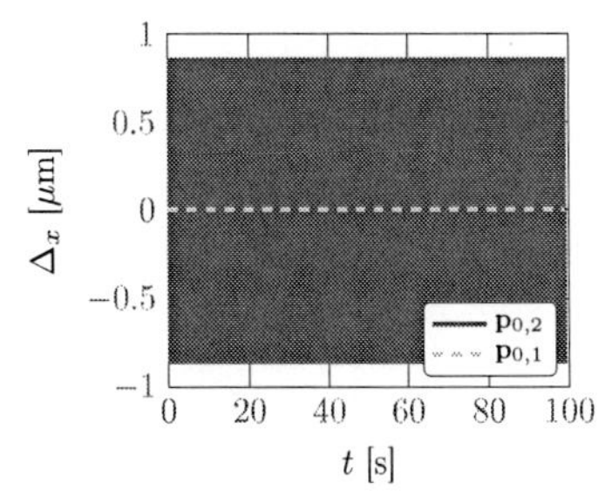

(a) Workpiece vibrations in x-direction (radial direction).

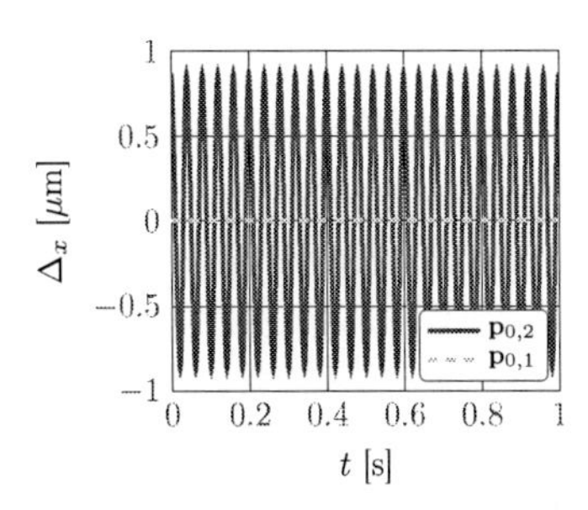

(b) Zoom to the first second of the vibrations of Figure 3.12a.

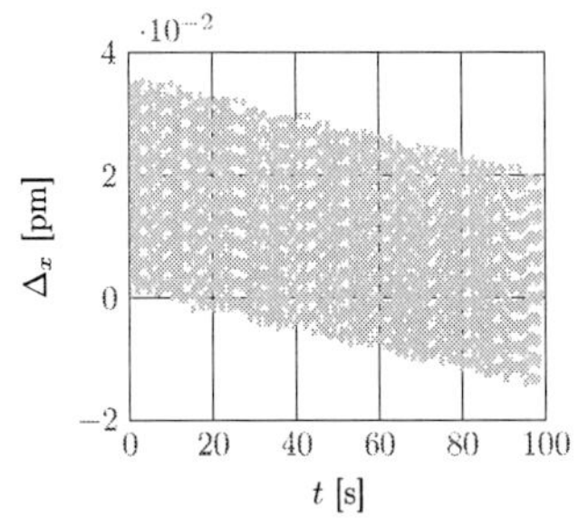

(c) Workpiece vibrations in x-direction for $\mathbf{p}_{0,1}$ (ideal case without any unbalance).

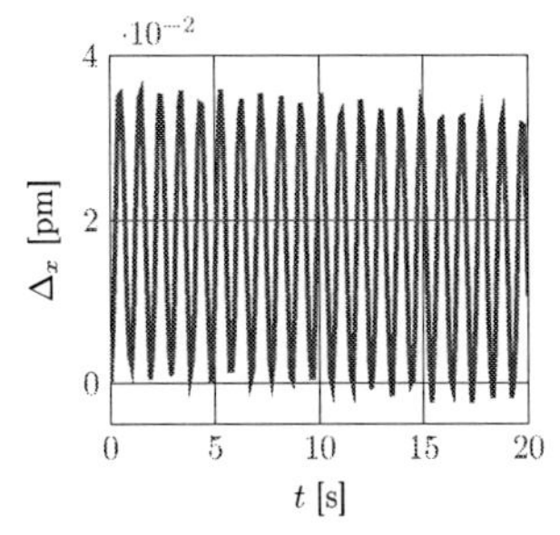

(d) Workpiece vibrations in x-direction for $\mathbf{p}_{0,1}$, zoom to the first seconds of Figure 3.12c.

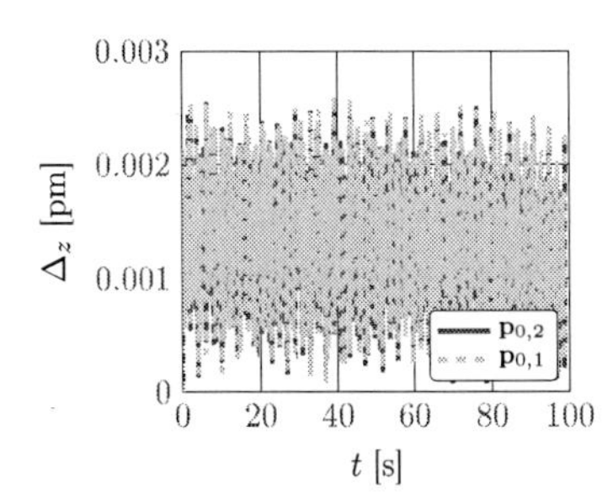

(e) Workpiece vibrations in z-direction (along the spindle axis).

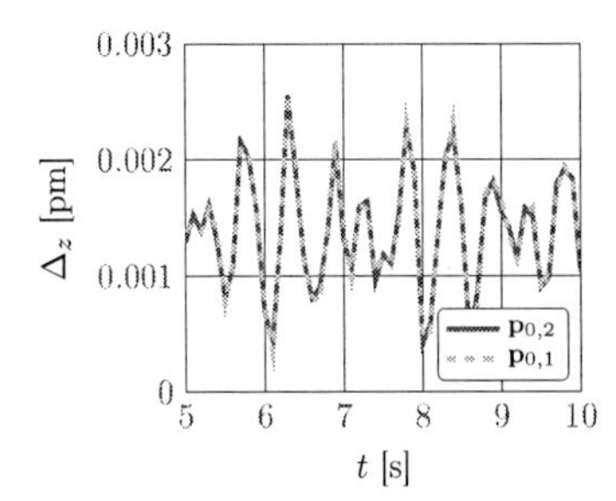

(f) Zoom to the workpiece vibrations in z-direction of Figure 3.12e.

Figure 3.12: Workpiece vibrations in radial and axial direction for two different unbalance distributions $\mathbf{p}_{0,1}$ for setting 1 (without unbalance) and $\mathbf{p}_{0,2}$ for setting 2 (mass at the workpiece). Note that one picometer is one part in a billion meter, i.e. $1\text{pm} = 10^{-12}\text{m} = 10^{-9}\text{mm}$.

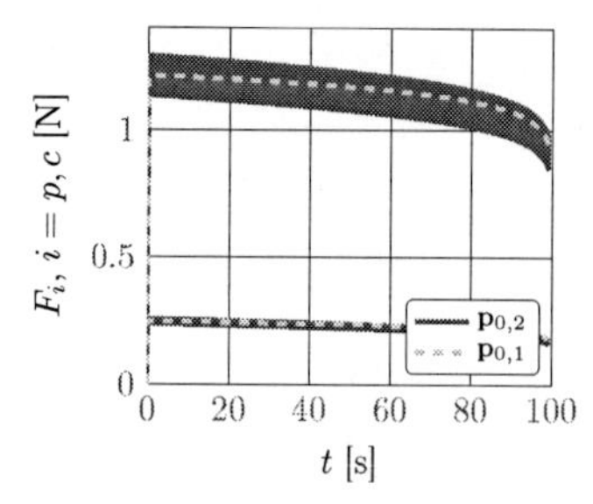

(a) Passive and cutting forces.

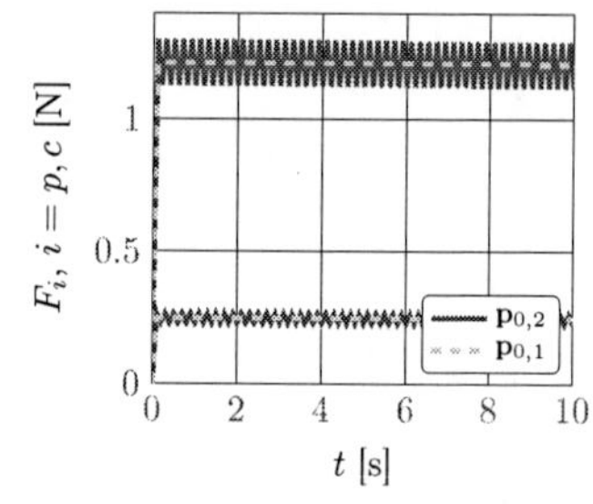

(b) Passive and cutting forces; zoom to the first ten seconds of Figure 3.13a.

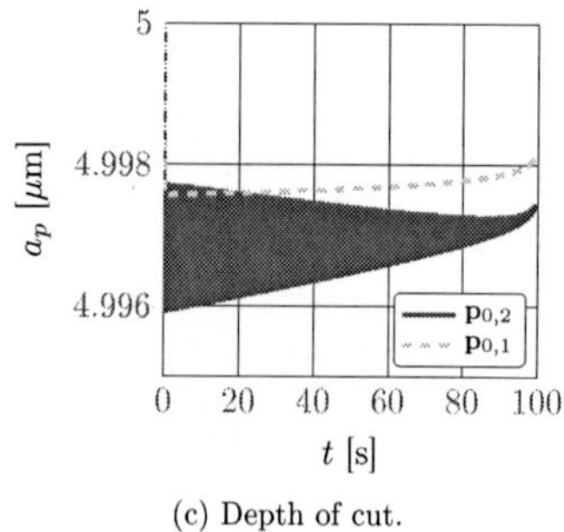

(c) Depth of cut.

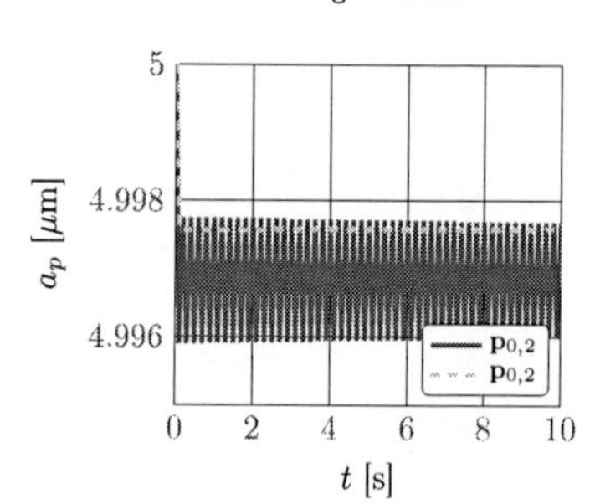

(d) Depth of cut; zoom to the first 10 seconds of Figure. 3.13c

Figure 3.13: Influence of unbalances on the cutting force components and depth of cut for two different unbalance distributions $\mathbf{p}_{0,1}$ (without unbalance) and $\mathbf{p}_{0,2}$ (mass at the workpiece).

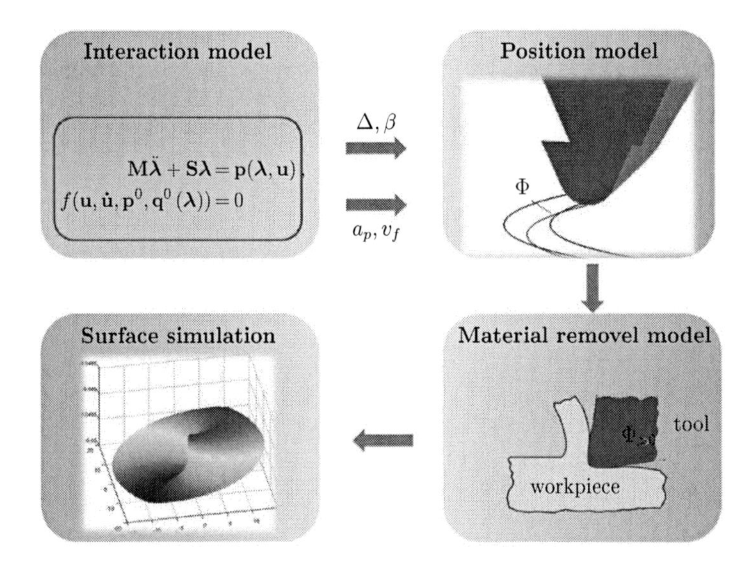

Figure 3.14: Approach in order to simulate the machined surface.

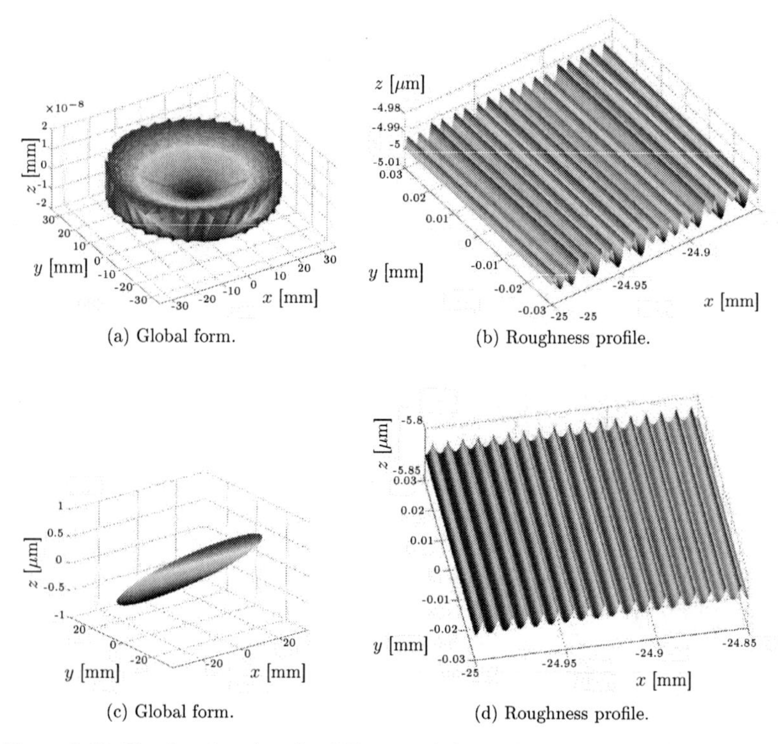

(a) Global form.

(b) Roughness profile.

(c) Global form.

(d) Roughness profile.

Figure 3.15: Simulated surface for different unbalance distributions: surface for the ideal case (no additional unbalance) and with unbalance, see setting 3 in Table 3.1.

Chapter 4

Identification of distributed parameters for the process model

In this chapter the determination of input parameters like depth of cut and feed velocity which are necessary for a prescribed precision of the output parameters, for example the tool path, is treated. The underlying forward operator is based on the ordinary differential equation system of the process model and is defined and analyzed in the following subsection. In Subsection 4.2 different applications of related inverse problems are solved with help of regularization methods with sparsity constraints. An example of parameter identification in case of the full coupled interaction model will be presented in Subsection 4.2.3.3.

4.1 Formulation of the problem

In the previous chapter, a system of ordinary differential equations describing the process model has been developed. For given input parameter for the process model and for the structure model, the solution contains information about the cutting forces and positions of the tool. The model has been simplified which results in the explicitly given system (ODE 2) of ordinary differential equations.

Neglecting the structural model, we obtain a system of differential equations of the form

$$\dot{\mathbf{u}} = f(\mathbf{u}, \mathbf{p}) \tag{4.1}$$

with right-hand side $f : D \times P \to E$,

$$(\mathbf{u}, \mathbf{p}) \mapsto (f_j(\mathbf{u}, \mathbf{p}))_{j=1}^{4}$$

and components

$$\begin{aligned}
f_1(\mathbf{u}, \mathbf{p}) &= -a_1 u_3 \,, && (4.2)\\
f_2(\mathbf{u}, \mathbf{p}) &= p_1 - a_7 (1 + a_2 u_4)(p_2 - u_3) \,, && (4.3)\\
f_3(\mathbf{u}, \mathbf{p}) &= a_{12} (p_2 - u_3) u_1^{\beta} u_2^{-a_5} u_3^{-a_4+1} - a_{11} u_3^2 u_1^{-1} \\
&\quad - a_6 u_3 u_2^{-1} \left[p_1 - a_7 (1 + a_2 u_4)(p_2 - u_3)\right] \,, && (4.4)\\
f_4(\mathbf{u}, \mathbf{p}) &= a_7 (p_2 - u_3) \,. && (4.5)
\end{aligned}$$

Hereby, we have set $D := (v_{c,\min}, v_{c,\max}) \times (a_{p,\min}, a_{p,\max}) \times (v_{x,\min}, v_{x,\max}) \times (0, \delta_{\max}) \subseteq \mathbb{R}_+^4$, $P := \left(a_{p,\min}^0, a_{p,\max}^0\right) \times (v_{f,\min}, v_{f,\max}) \subseteq \mathbb{R}_+^2$, $E := \mathbb{R}^4$, and $F := \mathbb{R}^2$. Thus,

$D \subset E$ open and $P \subset F$ open. The definition of D and P takes into account that the force model is validated for a certain parameter range and that the system is developed for positive variables.

The first observation concerns the differentiability of the function f like shown in the following lemma.

Lemma 45 (Differentiability of the right-hand side). *The right-hand side $f : D \times P \to E$ given by* (4.2)-(4.5) *of the simplified ordinary differential equation* (4.1) *is continuously differentiable, i.e. $f \in C^1(D \times P, E)$.*

Proof. Each component of the function f is a composition of continuous functions, and only the third component may have a discontinuity for $u_1 = u_2 = 0$ because of the negative exponents, but these points do not lie in the domain D of $f(\cdot, \mathbf{p})$. Therefore, f is continuous on $D \times P$.

We now compute the partial derivatives of each component f_i, $i = 1 \ldots 4$ with respect to $\mathbf{u}$ and $\mathbf{p}$, i.e. $\nabla_u f_i(\mathbf{u}, \mathbf{p}) = (\partial/\partial u_1 f_i(\mathbf{u}, \mathbf{p}), \ldots, \partial/\partial u_4 f_i(\mathbf{u}, \mathbf{p}))$ and $\nabla_p f_i(\mathbf{u}, \mathbf{p}) = (\partial/\partial p_1 f_i(\mathbf{u}, \mathbf{p}), \partial/\partial p_2 f_i(\mathbf{u}, \mathbf{p}))$. Differentiation of the first component yields

$$\nabla_u f_1(\mathbf{u}, \mathbf{p}) = (0, 0, -a_1, 0) \text{ and } \nabla_p f_1(\mathbf{u}, \mathbf{p}) = (0, 0) . \tag{4.6}$$

For the second component we obtain

$$\nabla_u f_2(\mathbf{u}, \mathbf{p}) = (0, 0, a_7(1 + a_2 u_4), -a_2 a_7 (p_2 - u_3)) \tag{4.7}$$

and

$$\nabla_p f_2(\mathbf{u}, \mathbf{p}) = (1, -a_7(1 + a_2 u_4)) , \tag{4.8}$$

and for the fourth component

$$\nabla_u f_4(\mathbf{u}, \mathbf{p}) = (0, 0, -a_7, 0) \tag{4.9}$$

as well as

$$\nabla_p f_4(\mathbf{u}, \mathbf{p}) = (0, a_7) . \tag{4.10}$$

For the remaining third component, we compute each partial derivative separately as follows

$$\begin{aligned}
\frac{\partial}{\partial u_1} f_3(\mathbf{u}, \mathbf{p}) &= a_{12}\beta (p_2 - u_3) u_1^{\beta-1} u_2^{-a_5} u_3^{1-a_4} + a_{11} u_3^2 u_1^{-2} , && (4.11) \\
\frac{\partial}{\partial u_2} f_3(\mathbf{u}, \mathbf{p}) &= -a_5 a_{12} (p_2 - u_3) u_1^{\beta} u_2^{-a_5-1} u_3^{1-a_4} \\
&\quad + a_6 u_3 u_2^{-2} [p_1 - a_7(1 + a_2 u_4)(p_2 - u_3)] , && (4.12) \\
\frac{\partial}{\partial u_3} f_3(\mathbf{u}, \mathbf{p}) &= -a_{12} u_1^{\beta} u_2^{-a_5} u_3^{-a_4+1} + a_{12}(1 - a_4)(p_2 - u_3) u_1^{\beta} u_2^{-a_5} u_3^{-a_4} \\
&\quad - 2a_{11} u_3 u_1^{-1} - a_6 u_2^{-1} [p_1 - a_7(1 + a_2 u_4)(p_2 - u_3)] \\
&\quad - a_6 a_7 u_3 u_2^{-1} (1 + a_2 u_4) , && (4.13) \\
\frac{\partial}{\partial u_4} f_3(\mathbf{u}, \mathbf{p}) &= a_2 a_6 a_7 u_3 u_2^{-1} (p_2 - u_3) , && (4.14) \\
\frac{\partial}{\partial p_1} f_3(\mathbf{u}, \mathbf{p}) &= -a_6 u_3 u_2^{-1} , && (4.15) \\
\frac{\partial}{\partial p_2} f_3(\mathbf{u}, \mathbf{p}) &= a_{12} u_1^{\beta} u_2^{-a_5} u_3^{-a_4+1} + a_6 a_7 u_3 u_2^{-1} (1 + a_2 u_4) . && (4.16)
\end{aligned}$$

Hence, the partial derivative of f with respect to $\mathbf{u}$ is given by

$$D_1 f(\mathbf{u},\mathbf{p}) = \begin{pmatrix} 0 & 0 & -a_1 & 0 \\ 0 & 0 & a_7(1+a_2u_4) & -a_2a_7(p_2-u_3) \\ \frac{\partial}{\partial u_1} f_3(\mathbf{u},\mathbf{p}) & \frac{\partial}{\partial u_2} f_3(\mathbf{u},\mathbf{p}) & \frac{\partial}{\partial u_3} f_3(\mathbf{u},\mathbf{p}) & \frac{\partial}{\partial u_4} f_3(\mathbf{u},\mathbf{p}) \\ 0 & 0 & -a_7 & 0 \end{pmatrix} \tag{4.17}$$

and with respect to $\mathbf{p}$ by

$$D_2 f(\mathbf{u},\mathbf{p}) = \begin{pmatrix} 0 & 0 \\ 1 & -a_7(1+a_2u_4) \\ -a_6u_3u_2^{-1} & a_{12}u_1^{\beta}u_2^{-a_5}u_3^{-a_4+1} + a_6a_7u_3u_2^{-1}(1+a_2u_4) \\ 0 & a_7 \end{pmatrix}. \tag{4.18}$$

Both derivatives $D_1 f(\mathbf{u},\mathbf{p})$ and $D_2 f(\mathbf{u},\mathbf{p})$ are continuous on $D \times P$ because the only critical points, where $u_i = 0$ for $i = 1, 2$, do not lie in the domain of f. Therefore, we have shown that f is continuously differentiable like claimed in the lemma. □

In the next subsection, we like to investigate the following parameter identification problem. Given an initial value problem with distributed parameter, we like to reconstruct the parameter from the solution of the IVP. We therefore introduce the so-called parameter-to-state map

$$\varphi : \mathrm{dom}(\varphi) \to U, \qquad \mathbf{p} \mapsto \mathbf{u}, \tag{4.19}$$

where $\mathbf{u}$ is the solution of the process model given as the IVP

$$\dot{\mathbf{u}} = f(\mathbf{u},\mathbf{p}), \qquad \mathbf{u}(t_0) = \mathbf{u}_0 \tag{4.20}$$

for a given parameter $\mathbf{p}$ and initial condition $\mathbf{u}_0$ at time t_0. The first question is whether the map is well-defined, i.e. whether a solution $\mathbf{u}$ exits for every $\mathbf{p}$ and under which conditions. The positive answer is given by the next lemma.

Lemma 46 (Unique solution of the process model)**.** *The simplified process model* (4.20)*, where the right hand side f is given by* (4.2)-(4.5)*, has an unique solution $\mathbf{u} : \mathcal{I}_u \to D$ on the interval $\mathcal{I}_u \subset \mathcal{I}$, which is relative open to $\mathcal{I} = [t_0, t_1]$, for every measurable essential bounded function $\mathbf{p} \in L^\infty(\mathcal{I}, P)$.*

Proof. Let the parameter $\mathbf{p} \in L^\infty(\mathcal{I}, P)$. According to Lemma 45, it holds that $f \in C^1(D \times P, E)$. Therefore, the assumptions of Theorem 27 about the existence of a solution of the parameter-dependent IVP are fulfilled, which completes the proof. □

Besides the well-posedness of the map φ, the lemma says that $\mathrm{dom}(\varphi)$ must be a subset of $L^\infty(\mathcal{I}, P)$. Moreovere, if $\mathcal{I}_u = \mathcal{I}$, then the parameter is called admissible. Furthermore, we know that the solution $\mathbf{u}$ of the IVP is an absolute continuous function, i.e. $\mathbf{u} \in AC(\mathcal{I}, D)$. Since D is n-dimensional, this is according to Theorem 6 equivalent to $\mathbf{u} \in W^{1,1}(\mathcal{I}, D)$. We can thus define the parameter-to-state-map. Therefore, remember the Definition 28, i.e.

$$D_\phi = \{(t_1, t_0, \mathbf{u}_0, \mathbf{p}) \,|\, t_0 < t_1, \mathbf{u}_0 \in D, \mathbf{p} : [t_0, t_1) \to P \text{ admissible for } \mathbf{u}_0\}.$$

Definition 47 (Parameter-to-state-map). The parameter-to-state-map is defined on the set

$$\operatorname{dom}(\varphi) = D_{t_1,t_0,\mathbf{u}_0} = \{\mathbf{p} | (t_1, t_0, \mathbf{u}_0, \mathbf{p}) \in D_\phi\} \subseteq L^\infty(I, P)$$

as the map

$$\begin{aligned} \varphi : D_{t,t_0,\mathbf{u}_0} &\to W^{1,1}(\mathcal{I}, D), \\ \mathbf{p} &\mapsto \mathbf{u}, \text{where } \mathbf{u} \text{ is a solution of (4.20).} \end{aligned}$$

Since in real life applications the solution $\mathbf{u}$ is not directly measurable, an observation operator B is introduced which maps the solution $\mathbf{u}$ to some measurable quantity, for example to the machined surface or force measurements. We define the linear bounded operator B as

$$\begin{aligned} B : W^{1,1}(\mathcal{I}, D) &\to \left(W^{1,1}(\mathcal{I}, \mathbb{R}) \times W^{1,1}(\mathcal{I}, \mathbb{R}) \times W^{1,1}(\mathcal{I}, \mathbb{R})\right), \\ \mathbf{u}(t) &\mapsto (\mathbf{x}(t), \mathbf{y}(t), \mathbf{z}(t)), \end{aligned} \tag{4.21}$$

where $\mathbf{x}$, $\mathbf{y}$ and $\mathbf{z}$ are the tool positions. They can be directly computed from the solution $\mathbf{u}$ of the IVP, i.e.

$$\begin{aligned} \mathbf{x} = -\frac{\mathbf{v}_c}{2\pi n} &= -\frac{\mathbf{u}_1}{a_1}, \\ \mathbf{y} = -C_y\delta &= -C_y\mathbf{u}_4, \\ \mathbf{z} = -\mathbf{a}_p &= -\mathbf{u}_2. \end{aligned}$$

The forward operator F is hence the composition of B and φ defined by

$$\begin{aligned} F : D_{t_1,t_0,\mathbf{u}_0} \subseteq L^\infty(\mathcal{I}, P) &\to \left(W^{1,1}(\mathcal{I}, \mathbb{R}) \times W^{1,1}(\mathcal{I}, \mathbb{R}) \times W^{1,1}(\mathcal{I}, \mathbb{R})\right), \\ \mathbf{p} &\mapsto (B \circ \varphi)(\mathbf{p}). \end{aligned} \tag{4.22}$$

Even in the case without any noise, the problem is ill-posed. Since the parameter-to-state-map maps every parameter to a solution with is absolutely continuous, it is possible to assume data which are less smooth, for example to assume a z-position which is not continuous but only piecewise continuous. Such data do not lie in the range of the forward operator.

In the next subsection we investigate the properties of the parameter-to-state-map and compute the derivative $\nabla\varphi$.

4.1.1 Properties of the parameter-to-state-map

The goal of this subsection is the analysis of the parameter-to-state map and hence of the forward operator. We start with the observation that the map φ is continuous.

Lemma 48. *The parameter-to-state-map φ is continuous, i.e.*

$$\varphi \in C\left(D_{t_1,t_0,\mathbf{u}_0}, W^{1,1}(\mathcal{I}, D)\right).$$

Proof. The continuity of the map follows immediately by Theorem 29 because the map φ is the restriction of the map ψ to the second argument for fixed initial condition $\mathbf{u}_0$, i.e. $\varphi : D_{t_1,t_0,\mathbf{u}_0} \to W^{1,1}(\mathcal{I}, D) \subset C^0(\mathcal{I}, D)$, $\varphi(\mathbf{p}) = \psi(\mathbf{u}_0, \mathbf{p})$. Since the right-hand side f of the IVP is continuously differentiable by Lemma 45, the assumptions of Theorem 29 are fulfilled, which concludes the proof. □

Lemma 49. *The parameter-to-state map φ is compact and hence completely continuous.*

Proof. Recalling the notion of compactness, we have to show that the image $M = \varphi(B)$ is relative compact for every bounded subset $B \subset D_{t_1,t_0,\mathbf{u}_0}$.

Thus, let B an arbitrary bounded subset of $D_{t_1,t_0,\mathbf{u}_0} \subseteq L^\infty(\mathcal{I}, P)$. The image of B is given by

$$M = \\ \{\mathbf{u} \in AC(\mathcal{I}, E) \,|\, \exists \mathbf{p} \in B \subset L^\infty(\mathcal{I}, P) : \dot{\mathbf{u}} = f(\mathbf{u}, \mathbf{p}) \text{ for a.a. } t \in \mathcal{I}, \mathbf{u}(t_0) = \mathbf{u}_0\}.$$

We begin by proving that M is bounded. Take a $\mathbf{u} \in M$ and estimate its norm via the integral representation (INT) like

$$\begin{aligned} \|\mathbf{u}(t)\| &\leq \|\mathbf{u}_0\| + \left\| \int_{t_0}^t f(\mathbf{u}(s), \mathbf{p}(s))\, \mathrm{d}s \right\|, \\ &\leq \|\mathbf{u}_0\| + \int_{t_0}^t \|f(\mathbf{u}(s), \mathbf{p}(s))\|\, \mathrm{d}s, \\ &\leq \|\mathbf{u}_0\| + \|f(\mathbf{u}, \mathbf{p})\|_\infty (t - t_0). \end{aligned}$$

Therefore, $\|\mathbf{u}\|_\infty \leq \|\mathbf{u}_0\| + \|f(\mathbf{u}, \mathbf{p})\|_\infty (t_1 - t_0) < \infty$, and hence $M = \varphi(B)$ is bounded.

We proceed to show that M is equicontinuous, i.e. we have to show that for every $t \in \mathcal{I}$ and arbitrary $\varepsilon > 0$ there is an open neighborhood $\mathcal{U} \subseteq \mathcal{I}$ of t such that

$$\|\mathbf{u}(t) - \mathbf{u}(s)\| \leq \varepsilon$$

for all $s \in \mathcal{U}$ and $\mathbf{u} \in M$. Choose $t \in \mathcal{I}$ and $\varepsilon > 0$ and take an arbitrary $\mathbf{u} \in M$. Define the constant C as the upper bound of

$$\sup_{\mathbf{u} \in M} \sup_{t \in \mathcal{I}} \|f(\mathbf{u}(t), \mathbf{p}(t))\|.$$

Set $\rho = \varepsilon / C$ and take the open ball $\mathcal{B}_\rho(t) \subset \mathcal{I}$ as the open neighborhood $\mathcal{U}$ of t in $\mathcal{I}$. Using again the representation (INT) of the solution of the IVP, we can estimate

$$\begin{aligned} \|\mathbf{u}(t) - \mathbf{u}(s)\| &= \left\| \int_s^t f(\mathbf{u}(\tau), \mathbf{p}(\tau))\, \mathrm{d}\tau \right\|, \\ &\leq \left| \int_s^t \|f(\mathbf{u}(\tau), \mathbf{p}(\tau))\|\, \mathrm{d}\tau \right|, \\ &\leq \|f(\mathbf{u}(\tau), \mathbf{p}(\tau))\|_\infty |t - s|, \\ &\leq C\rho = \varepsilon \end{aligned}$$

for all $s \in \mathcal{U} = \mathcal{B}_\rho(t)$. Thus, M is equicontinuous.

By the theorem of Arzéla-Ascoli (see Theorem 61 in the Appendix A), it follows that the set $M \subseteq W^{1,1}(\mathcal{I}, D) \subseteq C(\mathcal{I}, E)$ is relatively compact because $\mathcal{I}$ is a compact interval and E is an n-dimensional Banach space. Therefore, we have proven that φ maps bounded sets to relatively compact sets, and hence that it is a compact map.

Since the map φ is moreover continuous by the Lemma 48, it is completely continuous according to the Definition 40, and the proof is complete. □

Lemma 50. *The parameter-to-state map* φ *is differentiable, i.e.*

$$\varphi \in C^1\left(D_{t_1,t_0,\mathbf{u}_0}, W^{1,1}(\mathcal{I}, D)\right).$$

The derivative $\nabla\varphi(\mathbf{p})(\mathbf{q}) = \mathbf{v}$ *at* $\mathbf{p}$ *applied to* $\mathbf{q} \in L^\infty(\mathcal{I}, P)$ *is given by the solution* $\mathbf{v} : \mathcal{I} \to E$ *of the variational equation*

$$\dot{\mathbf{v}} = \frac{\partial}{\partial \mathbf{u}} f(\mathbf{u}, \mathbf{p})\,\mathbf{v} + \frac{\partial}{\partial \mathbf{p}} f(\mathbf{u}, \mathbf{p})\,\mathbf{q}, \tag{4.23}$$

$$\mathbf{v}(t_0) = 0. \tag{4.24}$$

Proof. This lemma is again a consequence of Lemma 45 which ensures that the assumptions of Theorem 29 are fulfilled. The assertion is then a direct consequence of Theorem 30. □

Remark 51. The derivative of the parameter-to-state-map is normally derived by the Implicit function theorem 58 in the theory of inverse parameter identification. Therefore, the problem is formulated as follows. The parameter-to-state map is defined as the map

$$\begin{aligned} \varphi : D_{t_1,t_0,\mathbf{u}_0} &\subseteq L^\infty(\mathcal{I}, P) \to W^{1,1}(\mathcal{I}, D), \\ \mathbf{p} &\mapsto \mathbf{u}, \end{aligned}$$

where $\mathbf{u}$ is the solution of the implicitly given equation

$$\Lambda(\mathbf{u}, \mathbf{p}, \mathbf{u}_0) = \mathbf{0} \tag{4.25}$$

with map $\Lambda : W^{1,1}(\mathcal{I}, D) \times L^\infty(\mathcal{I}, P) \times D \to C(\mathcal{I}, E) \times E$ defined by

$$\Lambda(\mathbf{u}(t), \mathbf{p}(t), \mathbf{u}_0) = \left(\dot{\mathbf{u}}(t) - f(\mathbf{u}(t), \mathbf{p}(t)),\ \mathbf{u}(t_0) = \mathbf{u}_0\right). \tag{4.26}$$

Since f is differentiable, we can compute the partial derivatives of Λ as

$$\begin{aligned} \frac{\partial \Lambda}{\partial \mathbf{u}}[\mathbf{u}, \mathbf{p}, \mathbf{u}_0](\mathbf{v}) &= \left(\dot{\mathbf{v}} - \frac{\partial}{\partial \mathbf{u}} f(\mathbf{u}, \mathbf{p})\,\mathbf{v},\ \mathbf{v}(t_0)\right), \\ \frac{\partial \Lambda}{\partial \mathbf{p}}[\mathbf{u}, \mathbf{p}, \mathbf{u}_0](\mathbf{q}) &= \left(-\frac{\partial}{\partial \mathbf{p}} f(\mathbf{u}, \mathbf{p})\,\mathbf{q},\ 0\right). \end{aligned}$$

By the Corollary 59 of the Implicit function theorem, it follows that

$$\frac{\partial \Lambda}{\partial \mathbf{u}}[\mathbf{u}, \mathbf{p}, \mathbf{u}_0](\nabla\varphi(\mathbf{p})\,\mathbf{q}) = -\frac{\partial \Lambda}{\partial \mathbf{p}}[\mathbf{u}, \mathbf{p}, \mathbf{u}_0]\,\mathbf{q}.$$

Consequently, the derivative $\nabla\varphi(\mathbf{p})\,\mathbf{q} = \mathbf{v}$ is given by

$$\begin{aligned} \dot{\mathbf{v}} - \frac{\partial}{\partial \mathbf{u}} f(\mathbf{u}, \mathbf{p})\,\mathbf{v} &= \frac{\partial}{\partial \mathbf{p}}(\mathbf{u}, \mathbf{p})\,\mathbf{q}, \\ \mathbf{v}(t_0) &= 0, \end{aligned}$$

which corresponds exactly to the variational equation (4.23)-(4.24) given in Lemma 50. This approach can also be applied to the implicitly given system of differential equations. For more information and further examples of parameter identification in distributed parameter systems, see [37, 47].

Lemma 52. *The forward operator is differentiable, i.e.*

$$F \in C^1 \left(D_{t_1,t_0,\mathbf{u}_0}, \left(W^{1,1}(\mathcal{I},\mathbb{R}) \times W^{1,1}(\mathcal{I},\mathbb{R}) \times W^{1,1}(\mathcal{I},\mathbb{R}) \right) \right).$$

Proof. Since the operator F is the composition of a differentiable operator and a linear bounded operator, the derivative can be determined by the chain rule

$$\nabla F(\mathbf{p}) = B \circ \nabla \varphi(\mathbf{p}) .$$

□

One possibility in order to invert the forward operator would be to compute the adjoined operator of the derivative $\nabla\varphi$ and hence of ∇F and to apply then for example a gradient method for the minimization of the corresponding regularized problem as explained Section 2.4 for the Hilbert space setting. In order to guarantee the existence of minimizer of the corresponding Tikhonov-functional

$$\|F(\mathbf{p}) - \mathbf{g}\|^2 + \alpha\Omega(\mathbf{p})$$

with a appropriate defined penalty term Ω, for example a Besov norm, it would be necessary to prove inter alia that φ is weakly sequentially closed.

We will not follow this approach because we will see in the next subsection that we can use the special structure of the differential equation system to deduce two linear inverse problems. However, for an overview about Tikhonov-regularization in Banach-spaces, we refer to [33, 53].

4.2 Identification of process parameters

In the following, we consider different inverse parameter identification problems for cutting processes and solve it by regularization methods with sparsity constraints. This will lead to optimized process parameters for feed rate and depth of cut. The following description follows the lines of [8].

4.2.1 Motivation

In this section we consider the problem of how to determine input parameters like depth of cut and feed velocity such that the output parameters, for example the tool position, obey a prescribed precision. We start with a example in order to illustrate the task. Assume that the depth of cut $a_p^0(t)$ and the feed velocity $v_f^0(t)$ are given as input parameters of the forward model F of the process model, see Figure 4.1. In Figure 4.2 the resulting tool tip positions $(x, z) = F(a_p^0, v_f^0)$ are shown. Due to deflections of the tool holder and changes of the feed velocity, the actual depth of cut $-z$ is reduced regarding the given input depth of cut a_p^0.

The task is the computation of new input parameters such that the resulting positions x and z cover exactly the solid lines in Figure 4.2. A simple optimization approach would

- start from heuristic choice of parameters,
- simulate the resulting output with these parameters,

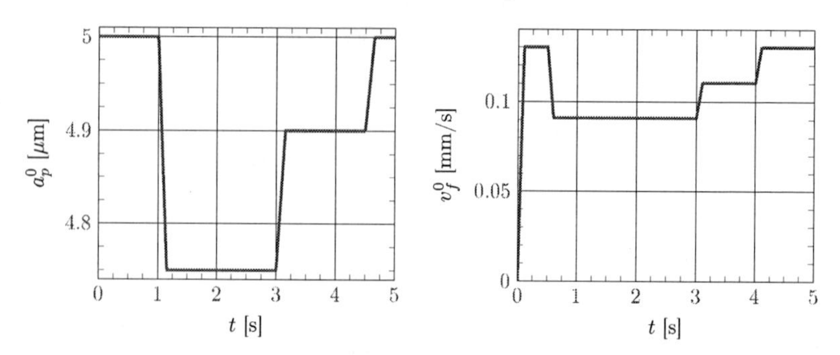

Figure 4.1: Example 1: Input depth of cut a_p^0 and input feed velocity v_f^0.

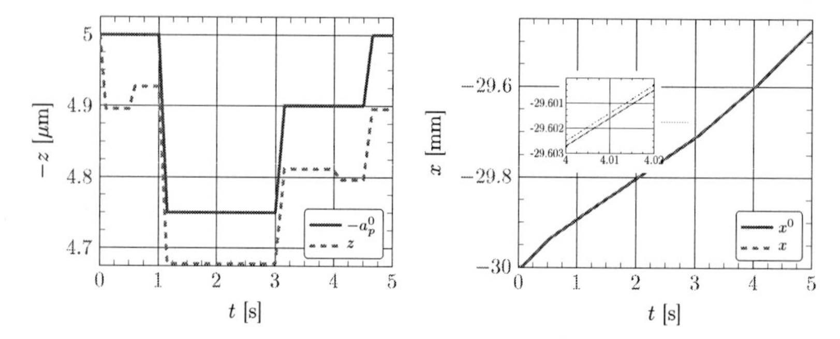

Figure 4.2: Example 1: Resulting position in z- and x-direction (dashed lines) for the input parameters a_p^0 and v_f^0 from Figure 4.1. The deflection of the tool and the changes in the feed cause the discrepancy regarding to the desired input position $z^0 = a_p^0$ and $x^0 = \int_0^t v_f^0(s)\,\mathrm{d}s$.

- determine the largest difference Δa_p^0 between the input depth of cut a_p^0 and the computed real position $-z$ (or take a mean value $\overline{\Delta a}_p$ over the whole time interval),

- correct the input parameter by $a_{p,corr} = a_p^0 + \Delta a_p^0$ or $a_{p,corr,2} = a_p^0 + \overline{\Delta a}_p$.

The position z_{corr} computed by the forward model $F(a_{p,corr}, v_f^0)$ should then be a better approximation to the desired tool tip position, see Figure 4.3a, but the desired position is still not achieved as shown in Figure 4.3b. In particular, it is not possible to reduce the steps caused by the jumps in the feed velocity without changing the input feed velocity. Nevertheless, we take this optimization procedure as a reference for our mathematical optimization strategy for the design problem presented in the Subsection 4.2.3. In order to solve the design problem with linear regularization methods, the inverse problem is reformulated in the next subsection.

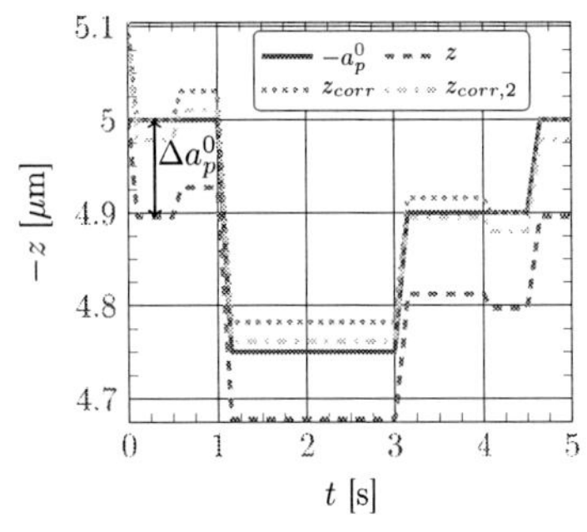

(a) Comparison of the corrected z-position and the desired position $z^* := -a_p^0$.

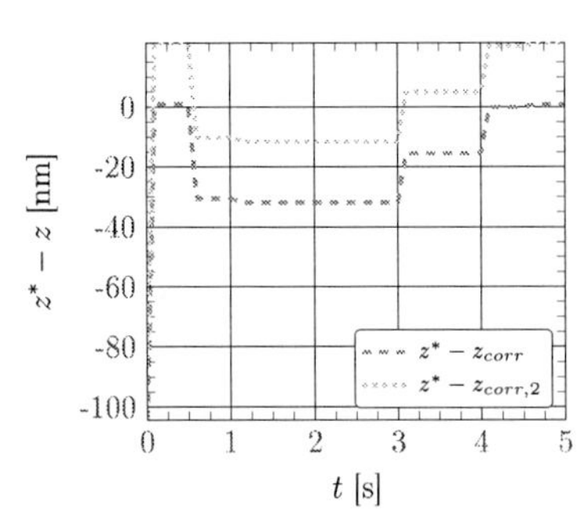

(b) Discrepancy between the computed correct positions and the desired position.

Figure 4.3: Example 1: The input depth of cut $a_p^0(t)$ plotted as solid line is taken as input for the forward model which gives the real position $-z(t)$. One simple correction in order to achieve the desired tool tip position $z^* = -a_p^0$ is to take $a_p^0(t) + \Delta a_p$ or $a_p^0(t) + \overline{\Delta a}_p$ as new input. The results z_{corr} and $z_{corr,2}$ are plotted as dotted and dotted dashed lines. The L_1-norms of the discrepancies are $\|z - z_{corr}\|_1 = 1.34\mu\text{m}$ and $\|z - z_{corr,2}\|_1 = 6.6\mu\text{m}$.

4.2.2 Reformulation of the inverse problem

The regularization techniques with sparsity constraints presented in Subsection 2.4.4 should be applied to the application of determining the best input parameters of the process model. We therefore assume that the positions $x(t)$ and $z(t)$ of the tool are given. By (3.9) and (3.11) it follows that

$$\int_0^t v_f^0(s)\,\mathrm{d}s = x(t) + r + \delta_x(t)$$

and

$$a_p^0(t) = -z(t) + \frac{1}{2l_h}\left(\delta_x^2(t) + \delta_y^2(t)\right) + \delta_z(t)\,.$$

From (3.20) and the formulas for the actual process parameter (3.15)- (3.17), we see that $\delta_i(t)$ $(i = x, y, z)$ can be formulated as a function of $x(t)$, $z(t)$ and $v_x(t) = \dot{x}(t)$. The deflection δ_z for example can be written as

$$\delta_z(t) = b_4\left(-z(t)\right)^{b_{10}}\dot{x}(t)^{b_7}\left(\alpha_{1,z}(-b_1 x(t))^{\beta_{1,z}} + \alpha_{2,z}(-b_1 x(t))^{-\beta_{2,z}}\right).$$

Thus, for given positions x and z we have first to compute the derivative $\dot{x}(t)$ of $x(t)$ and can then compute the right-hand sides of

$$\int v_f^0(s)\,\mathrm{d}s = x + r + \delta_x(x, z, \dot{x}) =: g_1 \tag{4.27}$$

and

$$a_p^0(t) = -z + \frac{b_{11}}{2}\left(\delta_x^2(x, z, \dot{x}) + \delta_y^2(x, z, \dot{x})\right) + \delta_z(x, z, \dot{x}) =: g_2\,, \tag{4.28}$$

which we label as g_1 and g_2. Both equations can be rewritten as an operator equation of our general setting (2.35), i.e.

$$\mathbf{A}_1(v_f^0) = g_1\,, \tag{4.29}$$

and

$$\mathbf{A}_2(a_p^0) = g_2\,, \tag{4.30}$$

where $\mathbf{A}_1$ is the integral operator

$$\begin{aligned} \mathbf{A}_1 : L^\infty\left(\mathcal{I},\mathbb{R}_+\right) &\rightarrow L^\infty\left(\mathcal{I},\mathbb{R}_+\right), \\ v &\mapsto \int_0^t v(s)\,\mathrm{d}s \end{aligned}$$

and $\mathbf{A}_2 : L^\infty\left(\mathcal{I},\mathbb{R}_+\right) \rightarrow L^\infty\left(\mathcal{I},\mathbb{R}_+\right)$ the identity.

Instead of inverting the nonlinear forward problem (4.22), solving the two linear equations in combination with the nonlinear algebraic equations for δ_i, $i = x, y, z$, is sufficient. In case of noisy data, the computation of the derivative of x^δ has of course to be regularized like explained in detail in Example 32 and 38. Since also the two linear compact operators $\mathbf{A}_1$ and $\mathbf{A}_2$ are ill-posed, the ill-posedness of the forward problem becomes obvious.

Therefore, the regularization methods with sparsity constraints proposed in subsection 2.4.4 have to be applied. Thus, the minimizing problems

$$v_{f,\alpha} = \min_{v_f} \|\mathbf{A}_1(v_f) - g_1\|_2^2 + \alpha\,\|v_f\|_1 \tag{4.31}$$

and

$$a_{p,\alpha} = \min_{a_p} \|\mathbf{A}_2(a_p) - g_2\|_2^2 + \alpha\,\|a_p\|_1\,. \tag{4.32}$$

have to be solved. Numerical examples for this approach will be presented in the next subsection.

Remark 53. Note that the RFSS-algorithm is developed for Hilbert spaces, but the searched parameters are assumed to be elements of the Banach space L^∞. Of course, L^∞ can be continuously embedded into the Hilbert space L^2, but then it is necessary to guarantee that the computed solutions of the Tikhonov-functional lie nevertheless in the parameter space of admissible parameters, i.e. in a subset of L^∞. Strictly speaking this theoretical gap has to be handled before applying the proposed algorithm. One possibility may be the application of a projection operator applied to the minimizer in order to force L^∞-solutions. An alternative may probably be a special choice of a L^∞-basis $\mathcal{B}$ in order to obtain correct solutions.

4.2.3 Parameter identification in ultra precision turning operations

In this subsection we will show different applications where the regularization method with sparsity constraints can be successfully applied. The first application, the design problem, covers the problem of computing the optimal input parameters in order to achieve a desired given tool tip position over time and is discussed in Subsection 4.2.3.1.

The second application concerns a similar task. This time not only desired positions are given but also initial input parameters. The goal is the computation of new input parameters by determining a sparse correction. This topic is treated in Subsection 4.2.3.2.

4.2.3.1 Design problem

The task for the design problem is the computation of optimal input parameters depth of cut a_p^0 and feed velocity v_f^0 for a given desired tool position $(x^*(t), z^*(t))$ under the additional assumption of a certain sparsity of the solution. The sparsity constraint is necessary because due to technical limitations of cutting machines arbitrary smooth input parameters are not possible. We therefore assume that the machine can realize inputs described by a set $\mathcal{B}$ of step functions or piecewise linear functions and ask the computed inputs to be sparse in exactly this function systems $\mathcal{B} = \{\varphi_i\}_{i\in\mathbb{N}}$. Consequently, we assume that only finite numbers n_a and n_v of coefficients are necessary in the representations

$$a_p^+ = \sum_{i=1}^{n_a} a_i\varphi_i \quad \text{and} \quad v_f^+ = \sum_{i=1}^{n_v} v_i\varphi_i \,.$$

To be able to use the RFSS algorithm 3.1, we introduce the linear operators

$$\begin{aligned} \mathbf{K}_1 : \quad & l^2 && \to L^\infty\left(\mathcal{I}, \mathbb{R}_+\right), \\ & \{v_i\}_i && \mapsto \mathbf{A}_1\left(\sum_{i\in\mathbb{N}} v_i\varphi_i\right) = \sum_{i\in\mathbb{N}} v_i\mathbf{A}_1\left(\varphi_i\right) \end{aligned} \tag{4.33}$$

and

$$\begin{aligned} \mathbf{K}_2 : \quad & l^2 && \to L^\infty\left(\mathcal{I}, \mathbb{R}_+\right), \\ & \{a_i\}_i && \mapsto \mathbf{A}_2\left(\sum_{i\in\mathbb{N}} a_i\varphi_i\right) = \sum_{i\in\mathbb{N}} a_i\varphi_i \,, \end{aligned} \tag{4.34}$$

which are acting directly on the coefficients. The two minimizing problems (4.31) and (4.32) are than equivalent to

$$v_{f,\alpha} = \min_{v_f} \|\mathbf{K}_1(v_f) - g_1\|_2^2 + \alpha\,\|v_f\|_{l^1} \tag{4.35}$$

and

$$a_{p,\alpha} = \min_{a_p} \|\mathbf{K}_2(a_p) - g_2\|_2^2 + \alpha\,\|a_p\|_{l^1} \,. \tag{4.36}$$

This formulation allows the introduction of the elastic net functional (2.54) and the use of the RFSS-algorithm for the computation of the minimizing solutions. Therefore, both operators are discretized - the operator $\mathbf{K}_2$ is than a matrix whose columns are vectors representing the functions $\varphi \in \mathcal{B}$ and $\mathbf{K}_1$ contains the integrated functions in its columns.

We first tested the approach with the example from Subsection 4.2.1.

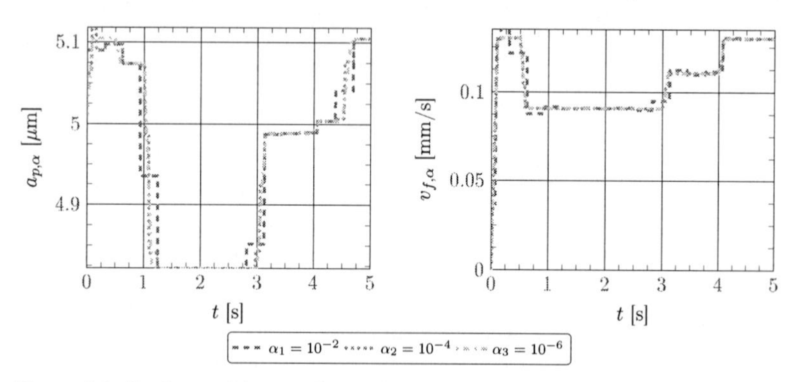

Figure 4.4: Design problem 1: Input depth of cut $a_{p,\alpha}$ and feed velocity $v_{f,\alpha}$ for three different regularization parameter $\alpha_1 = 10^{-2}$, $\alpha_2 = 10^{-4}$, and $\alpha_3 = 10^{-6}$. The solution for α_2 is less sparse but has less jumps.

Example 54 (Design problem 1)**.** The given desired position $z^*(t)$ is plotted as a solid line in Figure 4.5 and is used together with the desired position $x^*(t)$ to determine the right hand sides g_1 and g_2 using (4.27) and (4.28). We then compute the input parameter $a_{p,\alpha}$ and $v_{f,a}$ for different regularization parameters α by solving the minimizing problems (4.35) and (4.36). The determined inputs are shown in Figure 4.4.

The solution $a_{p,\alpha}$ for $\alpha_1 = 10^{-2}$ has 8 non-zero coefficients in its expansion, wheres the solutions for $\alpha_2 = 10^{-4}$ and $\alpha_3 = 10^{-6}$ consists of 17 and 108 basis elements. A similar effect can be observed for $v_{f,\alpha}$, where 11, 34, and 74 coefficients are needed in the basis expansion. As expected, the solution for the biggest regularization parameter ($\alpha_1 = 10^{-2}$) is sparser than the other solutions but has bigger jumps.

We put the input parameters into our forward model in order to compare the resulting positions $x_\alpha(t)$ and $z_\alpha(t)$ with the given desired ones. In Figure 4.5 we can see the position z_α for different regularization parameters and the given positions z^*. The solution for $\alpha_3 = 10^{-6}$ covers very well the given position, see also the discrepancy between z^* and z_α in Figure 4.6. The highest deviations can be observed at the step positions of the input parameters.

The norms of the differences $x^* - x_\alpha$ and $z^* - z_\alpha$ are collected in Table 4.1 and compared to the discrepancies without any correction and that ones of the simple correction method of Subsection 4.2.1, namely with the correction z_{corr} of taking the biggest discrepancy and the correction $z_{corr,2}$ of taking the mean value Δz. Clearly, the regularization method provides better results than the simple correction method.

4.2.3.2 Determination of a sparse correction

The goal of this subsection is similar to that one of the last one. Again the desired positions x^* and y^* are given, but additionally we assume that we have initial input parameters v_f^0 and a_p^0. We now are looking for optimal input parameters v_f and a_p

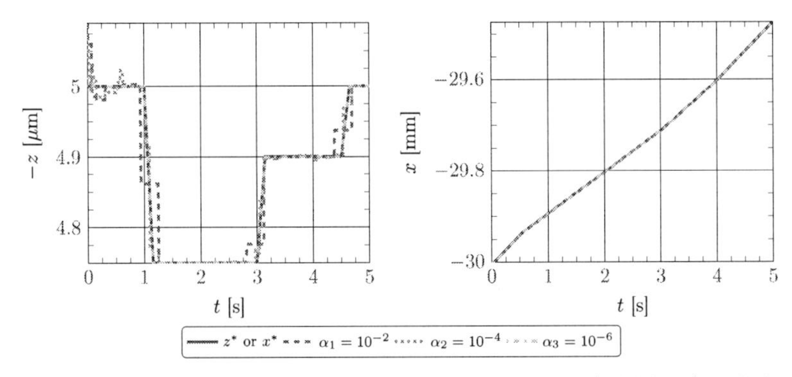

Figure 4.5: Design problem 1: Desired positions z^* and x^*(solid lines) and the corrected resulting positions $(z_\alpha, x_\alpha) = F(a_{p,\alpha}, v_{f,\alpha})$ for different regularization parameters $\alpha_1 = 10^{-2}$, $\alpha_2 = 10^{-4}$, and $\alpha_3 = 10^{-6}$.

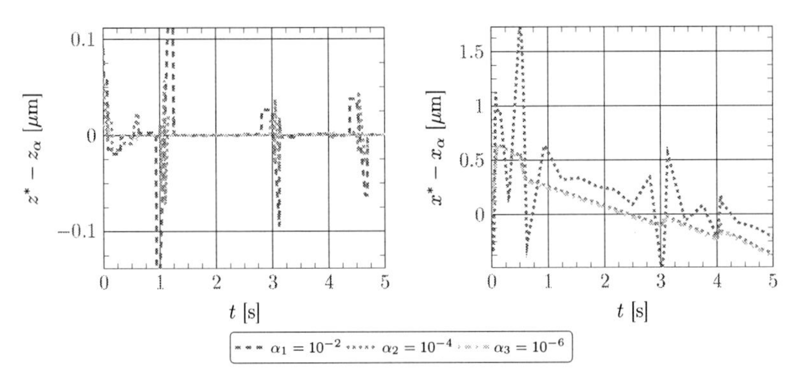

Figure 4.6: The difference between the desired position z^* and the resulting positions z_α as well as between x^* and x_α.

such that the correction to the initial ones is sparse, i.e. we are searching

$$v_f = v_f^0 + \Delta v_f$$

and

$$a_p = a_p^0 + \Delta a_p$$

with Δv_f and Δa_p sparse, such that

$$\mathbf{A}_1(v_f) = g_1(x^*, z^*, \dot{x}^*)$$

and

$$\mathbf{A}_2(a_p) = g_2(x^*, z^*, \dot{x}^*) \ .$$

We want the corrections to be sparse because we wish to modify the initial input parameters only at as few points as possible. We can use the same formulation in

Table 4.1: Design problem 1: The norms of the discrepancy [μm].

$\bar{z}$	$\|z^* - \bar{z}\|_2$	$\|z^* - \bar{z}\|_1$	$\|z^* - \bar{z}\|_\infty$
z	1.925	42.874	0.104
z_{corr}	0.567	10.334	0.104
$z_{corr,2}$	0.343	6.663	0.084
$z_{\alpha=10^{-2}}$	0.688	6.451	0.139
$z_{\alpha=10^{-4}}$	0.254	2.462	0.072
$z_{\alpha=10^{-6}}$	0.035	0.212	0.010
$\bar{x}$	$\|x^* - \bar{x}\|_2$	$\|x^* - \bar{x}\|_1$	$\|x^* - \bar{x}\|_\infty$
x	5.716	103.733	0.620
$x_{\alpha=10^{-2}}$	9.184	145.485	1.727
$x_{\alpha=10^{-4}}$	5.786	103.266	0.641
$x_{\alpha=10^{-6}}$	5.715	103.748	0.620

order to solve the problem as in the last subsection because the two operators $\mathbf{A}_1$ and $\mathbf{A}_2$ are linear. Thus, we have to solve the operator equations

$$\mathbf{A}_1(\Delta v_f) = g_1(x^*, y^*, \dot{y}^*) - \mathbf{A}_1(v_f^0) =: \tilde{g}_1 \tag{4.37}$$

and

$$\mathbf{A}_2(\Delta a_p) = g_2(x^*, y^*, \dot{y}^*) - \mathbf{A}_2(a_p^0) =: \tilde{g}_2 \,. \tag{4.38}$$

We come back to the example of Section 4.2.1.

Example 55 (Sparse correction problem 1)**.** The desired non-constant input parameters a_p^0 and v_f^0 are plotted as solid lines in Figure 4.7. If these parameters are plugged into the forward model, the position z is obtained, see the solid line in Figure 4.8a. We now wish to achieve a slightly modified position z^* whereas we do not want to change $x = x^*$.

The computed correction $\Delta a_{p,\alpha}$ for three different α is illustrated in Figure 4.8c. The corrected inputs $a_{p,\alpha} = a_p^0 + \Delta a_{p,\alpha}$ are plotted in Figure 4.8d. Together with the new input feed velocities $v_{f,\alpha}$ in Figure 4.9a, the new positions $(x_\alpha, z_\alpha) = F(a_{p,\alpha}, v_{f,\alpha})$ are computed. The results for the z and x-position are shown in Figure 4.8b and 4.9b.

Like in the previous example of the design problem, the solution for the smallest regularization parameter $\alpha_3 = 10^{-6}$ covers the desired position best. It is the solution with most non-zero coefficients, namely 51 instead of 23 and 2 for $\alpha_2 = 10^{-5}$ and $\alpha_1 = 10^{-4}$. The same result is reflected in the discrepancy between desired position and computed position in Figure 4.10 as well as by their norms in Table 4.2. The desired position z^* is well approximated by the corrected position z_α.

The next example is a modification of the last one. This time, we do not change the position z but x, i.e. we wish to achieve a position x^* such that the necessary input feed velocity v_f is a small modification of the given desired input feed velocity v_f^0.

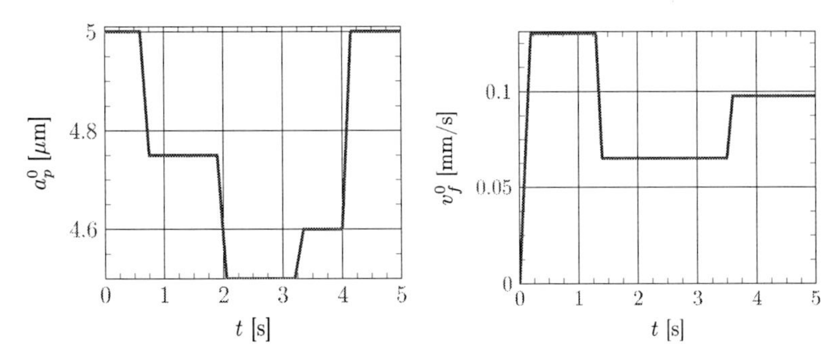

Figure 4.7: Sparse correction problem 1: Input parameters a_p^0 and v_f^0.

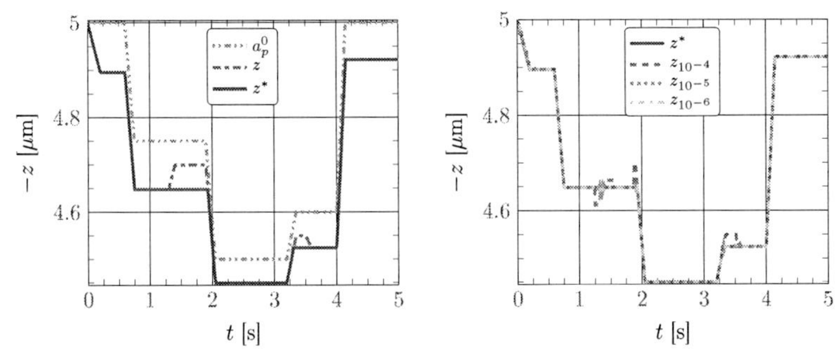

(a) Actual position z resulting from the depth of cut a_p^0 and desired new position z^*.

(b) Corrected z_α-positions computed with the new input depth of cut $a_{p,\alpha}$.

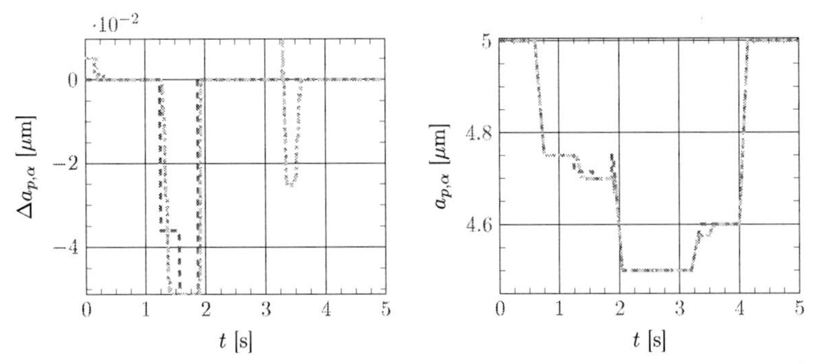

(c) Sparse correction $\Delta a_{p,\alpha}$ for different regularization parameters α.

(d) Computed new depth of cut $a_{p,\alpha} = a_p^0 + \Delta a_{p,\alpha}$.

$\alpha_1 = 10^{-4}$ $\alpha_2 = 10^{-5}$ $\alpha_3 = 10^{-6}$

Figure 4.8: Sparse correction problem 1.

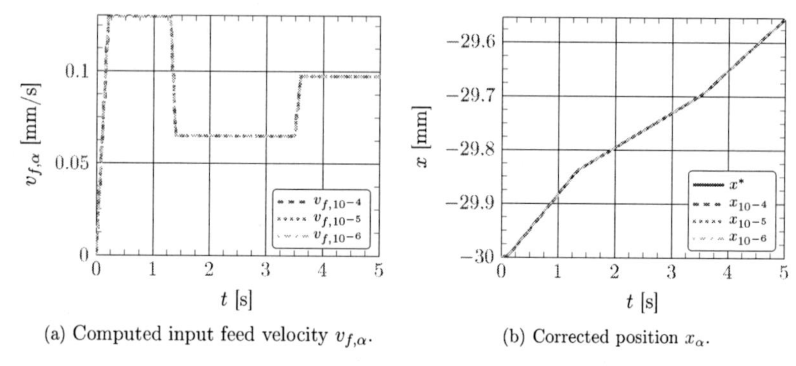

(a) Computed input feed velocity $v_{f,\alpha}$. (b) Corrected position x_α.

Figure 4.9: Sparse correction problem 1: Computed new input feed velocity $v_{f,\alpha}$ and resulting x-position x_α for different regularization parameter α.

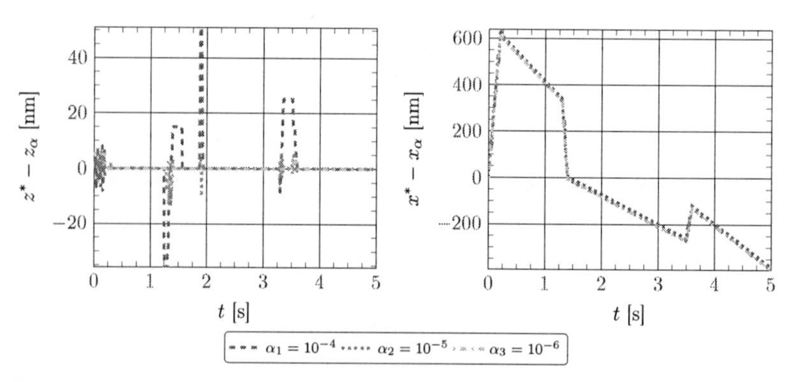

Figure 4.10: Difference between desired positions x^* and z^* and the computed positions $(x_\alpha, z_\alpha) = F(a_{p,\alpha}, v_{f,\alpha})$. The L_1-norm is $\|z - z_\alpha\|_1 = 186$nm for $\alpha_3 = 10^{-6}$.

Table 4.2: Sparse correction problem 1: The norms of the discrepancy without and with sparse correction of the position z.

$\bar{z}$	$\|z^* - \bar{z}\|_2$	$\|z^* - \bar{z}\|_1$	$\|z^* - \bar{z}\|_\infty$	[nm]
z	397	3524	51	
$z_{\alpha=10^{-4}}$	192	1479	51	
$z_{\alpha=10^{-5}}$	37	253	13	
$z_{\alpha=10^{-6}}$	29	186	5	
$\bar{x}$	$\|x^* - \bar{x}\|_2$	$\|x^* - \bar{x}\|_1$	$\|x^* - \bar{x}\|_\infty$	[μm]
x	6.73	130.64	0.608	
$x_{\alpha=10^{-4}}$	6.72	127.67	0.639	
$x_{\alpha=10^{-5}}$	6.72	130.67	0.608	
$x_{\alpha=10^{-6}}$	6.72	130.67	0.608	

Example 56 (Sparse correction problem 2). The point of departure is illustrated in Figure 4.11. Instead of the position x^0 which would be the result of the input feed velocity v_f^0, the position x^* should be reached, see Figure 4.11b. This position corresponds to a new feed velocity v_f^* which is a slightly modified version of the initial feed velocity v_f^0, see Figure 4.12a. The z-position is not modified, i.e. $z^* = z$.

The computed corrections $\Delta v_{f,\alpha}$ for $\alpha_1 = 10^{-4}$ and $\alpha_2 = 10^{-6}$ are illustrated in Figure 4.12c and are clearly sparse. It can be seen that the correction concerns only the small regions where x^* differs from x^0, compare with Figure 4.12c and 4.12a. The corrected input feed rate $v_{f,\alpha} = v_f^0 + \Delta v_{f,\alpha}$ as well as the computed input depth of cut are plotted in Figure 4.12d and 4.13. The resulting positions for $(x_\alpha, z_\alpha) = F(a_{p,\alpha}, v_{f,\alpha})$ are plotted as dashed and dotted lines in Figures 4.12b and 4.13. Both solutions for $\alpha_1 = 10^{-4}$ and $\alpha_2 = 10^{-6}$ provide similar results for the resulting x-position, whereas the z-position is again better approximated for the smaller regularization parameter.

In Table 4.3 the norms of their differences between x^* and x_α as well as z^* and z_α are collected. The decreasing discrepancies confirm the visual impression for the quality of approximation and the best parameter choice.

Table 4.3: Sparse correction problem2: The norms of the deviation without and with sparse correction.

$\bar{z}$	$\|z^* - \bar{z}\|_2$	$\|z^* - \bar{z}\|_1$	$\|z^* - \bar{z}\|_\infty$	[nm]
z	1670	36093	104	
$z_{\alpha=10^{-4}}$	282	3051	57	
$z_{\alpha=10^{-5}}$	41	440	10	
$z_{\alpha=10^{-6}}$	26	211	5	
$\bar{x}$	$\|x^* - \bar{x}\|_2$	$\|x^* - \bar{x}\|_1$	$\|x^* - \bar{x}\|_\infty$	[μm]
x	93.5	487.5	32.0	
$x_{\alpha=10^{-4}}$	26.6	183.4	8.0	
$x_{\alpha=10^{-5}}$	17.1	117.1	3.2	
$x_{\alpha=10^{-6}}$	16.3	113.9	3.2	

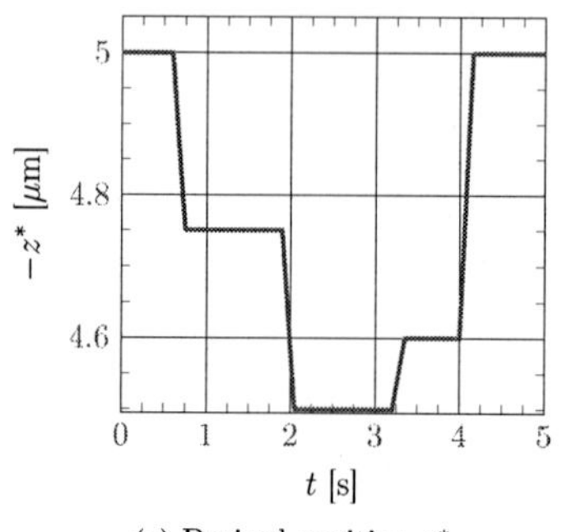

(a) Desired position z^*

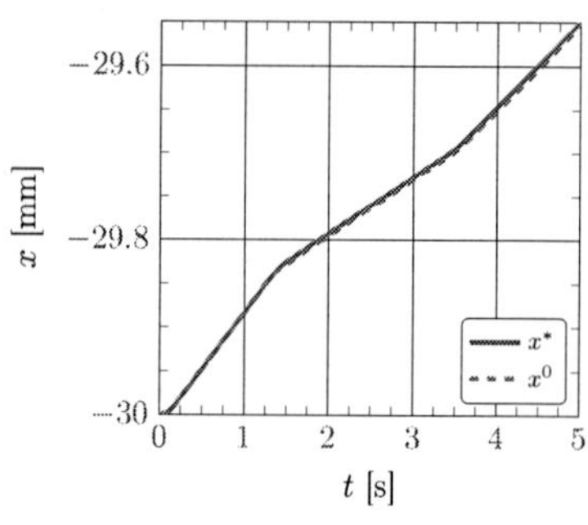

(b) Desired position x^* and input position x^0 corresponding to v_f^0.

Figure 4.11: Sparse correction problem 2: Desired x- and z-position.

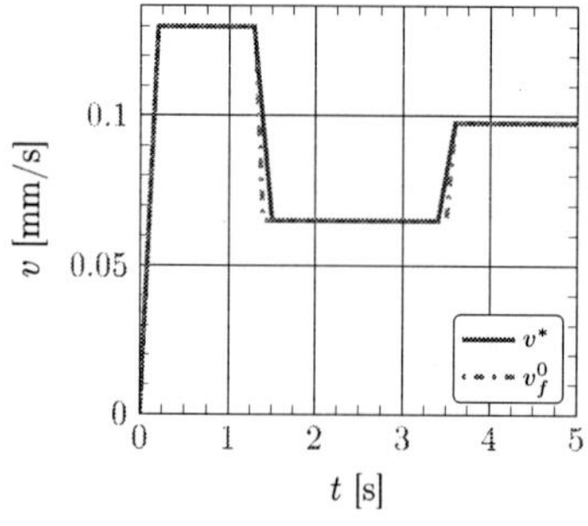

(a) Input velocity v_f^0 and desired velocity v_f^* computed from x^*.

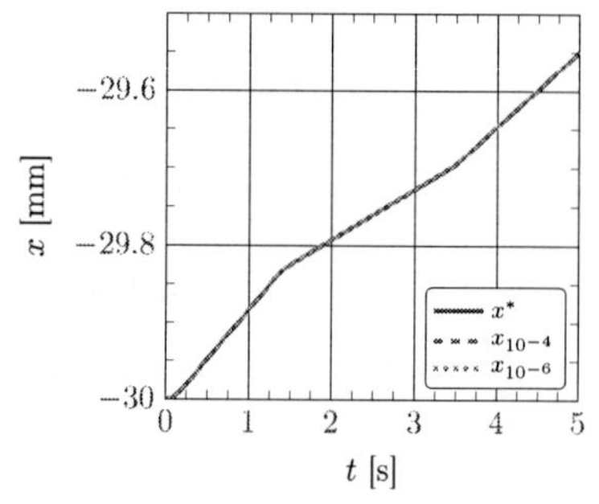

(b) Desired position x^* and computed position x_α.

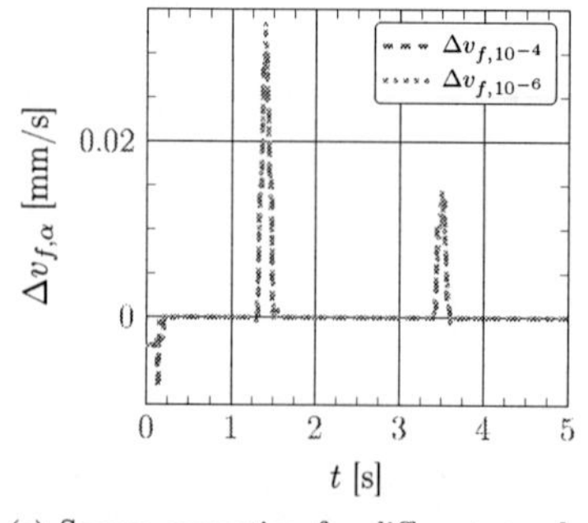

(c) Sparse correction for different regularization parameters α.

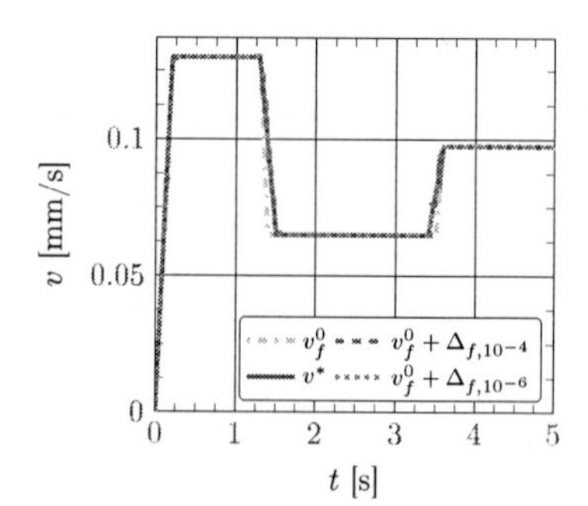

(d) Input, desired and computed new input feed velocities $v_{f,\alpha}$.

Figure 4.12: Example of a sparse correction $\Delta v_{f,\alpha}$ for the input feed velocity v_f^0 in order to archive a desired position x^*.

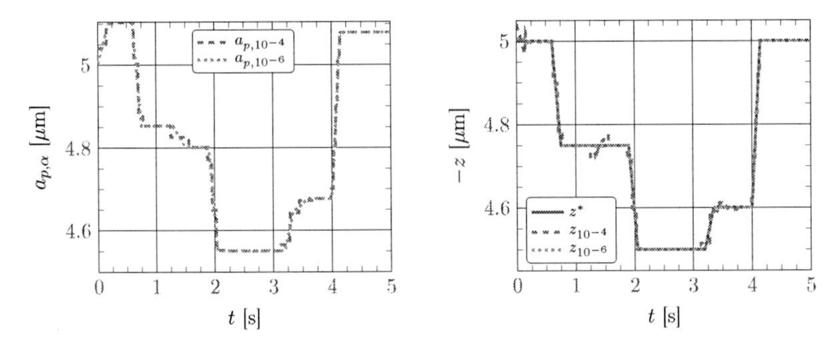

Figure 4.13: Computed input depth of cut $a_{p,\alpha}$ and corresponding position z_α, both for two different regularization parameters α.

Both example for the computation of sparse corrections show that the application of regularization methods with L_1-penalty provide indeed sparse solutions and good approximations to the desired tool positions.

4.2.3.3 Parameter identification under consideration of machine vibrations

In the previous subsections, two different possible applications - the design and the sparse correction problem - are treated. Both are based on the inversion of the process model, and the influence of machine vibrations and their impact on the process machine interaction are neglected. In this subsection the complete forward model (3.62)-(3.63) including the process machine interaction is used, i.e. the forward model is described by

$$
\begin{aligned}
\mathbf{M}\ddot{\boldsymbol{\lambda}}(t) + \mathbf{S}\boldsymbol{\lambda}(t) &= \mathbf{p}_{unb}(t) + \mathbf{p}_{cut}(t, \boldsymbol{\lambda}(t), \mathbf{u}(t)), && (4.39)\\
f(\mathbf{u}(t), \dot{\mathbf{u}}(t), \mathbf{p}^0(t), \mathbf{q}^0(t, \boldsymbol{\lambda}(t))) &= 0. && (4.40)
\end{aligned}
$$

Despite of $\mathbf{p}^0(t) = (v_f^0(t), a_p^0(t))$, the unbalance distribution $\mathbf{p}_{unb}(t)$ is needed as input parameter . For given mass vector $\boldsymbol{\Delta}\mathbf{m} \in \mathbb{R}_+^N$ and angular positions ϕ the unbalance distribution $\mathbf{p}_{unb} \in C^\infty\left(\mathcal{I}, \mathbb{C}^N\right)$ is determined. In the following, only the $\mathbf{p}^0$ is searched because $\mathbf{p}_{unb}$ is assumed to be known. Consequently, the parameter-to-state map is defined on the set

$$D_{t,t_0,\mathbf{u}_0} \subseteq L^\infty(\mathcal{I}, P)$$

as the map

$$
\begin{aligned}
\Psi : D_{t,t_0,\mathbf{u}_0} &\to W^{1,1}(\mathcal{I}, D) \times C^1(\mathcal{I}, D_2)\\
\mathbf{p} &\mapsto \nu = (\mathbf{u}, \boldsymbol{\lambda}) \text{ ,where } \nu \text{ is a solution of } (4.39) - (4.40).
\end{aligned}
$$

The observation operator $\tilde{B}$ acts only on the first part $\mathbf{u}$ of the solution ν, where it is identical to the observation operator B, see (4.21), i.e. $\tilde{B}|_{W^{1,1}} \equiv B$. Therefore, the forward operator is given by

$$
\begin{aligned}
\tilde{F} : D_{t,t_0,\mathbf{u}_0} \subseteq L^\infty(\mathcal{I}, P) &\to \left(W^{1,1}(\mathcal{I}, \mathbb{R}) \times W^{1,1}(\mathcal{I}, \mathbb{R}) \times W^{1,1}(\mathcal{I}, \mathbb{R})\right),\\
\mathbf{p} &\mapsto \left(\tilde{B} \circ \Psi\right)(\mathbf{p}).
\end{aligned}
$$

Nevertheless, the same reformulation of the inverse problem like before can be used. Therefore, assume that x and z are known. Recall the equations (3.12)-(3.14) for the x and z-position and observe that

$$\int_0^t v_f^0(s)\,\mathrm{d}s = x(t) + r + \delta_x(x,z,\dot{x}) + \Delta_x(t)$$

and

$$a_p^0(t) = -z(t) + \frac{b_{11}}{2}\left(\delta_x^2(x,z,\dot{x}) + \delta_y^2(x,z,\dot{x})\right) + \delta_z(x,z,\dot{x}) \\ - \Delta_z(t) + x(t)\tan\left(\beta_y(t)\right).$$

The expressions $\eta_1 = \Delta_x$ and $\eta_2 = -\Delta_z + x\tan(\beta_y)$ describing the influence of the machine vibrations can be regarded as some noise or disturbance to the given positions x and z. Thus, labeling these disturbed positions as $x^\eta = x + \eta_1$ and $z^\eta = z + \eta_2$, we obtain

$$\int_0^t v_f^0(s)\,\mathrm{d}s = x^\eta(t) + r + \delta_x(x,z,\dot{x}) =: g_1^\eta \tag{4.41}$$

and

$$a_p^0(t) = -z^\eta(t) + \frac{b_{11}}{2}\left(\delta_x^2(x,z,\dot{x}) + \delta_y^2(x,z,\dot{x})\right) + \delta_z(x,z,\dot{x}) =: g_2^\eta . \tag{4.42}$$

Like before, the two operator equations

$$\mathbf{A}_1(v_f) = g_1^\eta$$

and

$$\mathbf{A}_2(a_p) = g_2^\eta$$

with the integral operator $\mathbf{A}_1$ and the identity $\mathbf{A}_2$ have to be solved. The solution is again computed by introducing the linear operator $\mathbf{K}_1$ and $\mathbf{K}_2$ and solving the minimizing problems

$$v_{f,\alpha} = \min_{v_f} \|\mathbf{K}_1(v_f) - g_1^\eta\|_2^2 + \alpha\,\|v_f\|_{l^1} \tag{4.43}$$

and

$$a_{p,\alpha} = \min_{a_p} \|\mathbf{K}_2(a_p) - g_2^\eta\|_2^2 + \alpha\,\|a_p\|_{l^1} . \tag{4.44}$$

For an illustration of the procedure, we come back to the Example 54 for the design problem.

Example 57 (Design problem with machine vibrations). The same input parameter v_f^0 and a_p^0 like in Example 54, Figure 4.1, are used. The unbalance distribution in the machine model is set to zero, i.e. $\mathbf{p}_{unb}(t) = 0$ for all t. Consequently, only process induced machine vibrations are considered. The resulting positions and x- and z-direction are shown in Figure 4.14 and are compared to the corresponding input positions x^0 and z^0. They are disturbed by the process induced vibrations.

As before, the deflections and thus the right hand sides g_1^η and g_2^η are computed by x^η and z^η using (4.41) and (4.42), and the minimizing problems (4.43) and (4.44) are

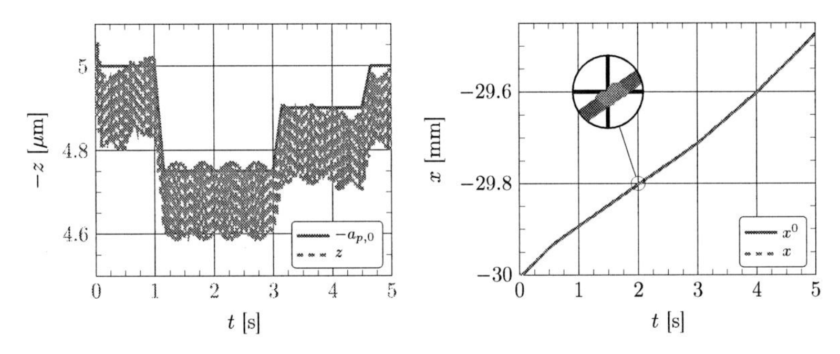

Figure 4.14: The z- and x-position computed for the input parameter a_p^0 and v_f^0 of the motivating example (see Figure 4.1) by $(x, z) = \tilde{F}\left(v_f^0, a_p^0\right)$.

solved. The optimal input depth of cut $a_{p,\alpha}$ for a regularization parameter $\alpha = 10^{-6}$ is plotted in Figure 4.15a as a dotted line, and the corresponding z-position z_α is shown in Figure 4.15b together with the desired position $z^* = -a_p^0$ and the given position z (dashed line). Because of the vibrations, the quality of the reconstruction is difficult to evaluate. Therefore, the results of the forward model z and z_α are filtered with help of the fast wavelet transformation and compared again with z^* in Figure 4.15d. This clearly shows the high improvement of the computed position z_α which approximates the desired position very well.

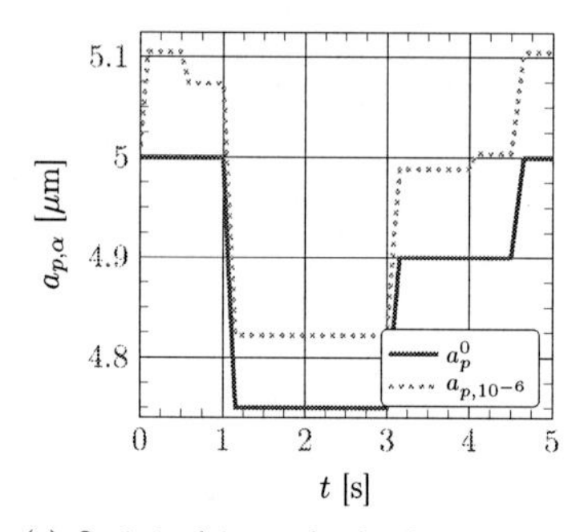

(a) Optimized input depth of cut $a_{p,\alpha}$ and initial input depth of cut a_p^0.

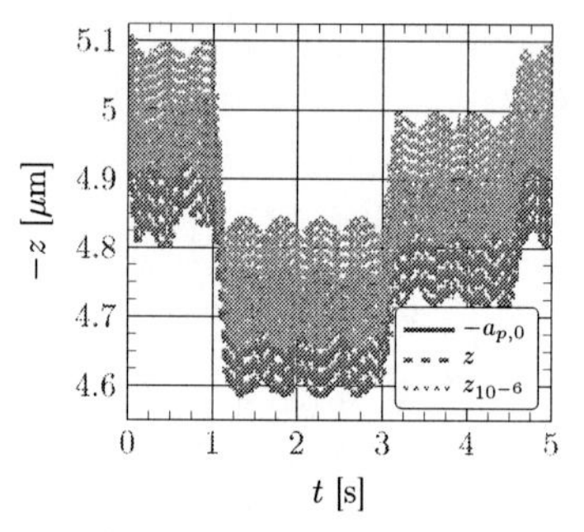

(b) Resulting positions z and z_α.

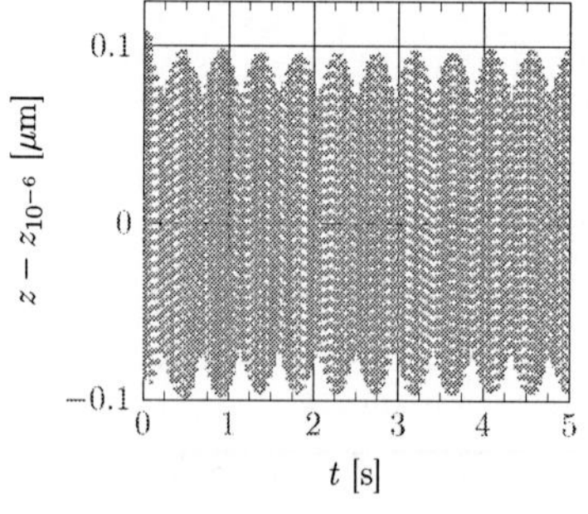

(c) Difference between computed position z_α and desired position $z^* = a_p^0$.

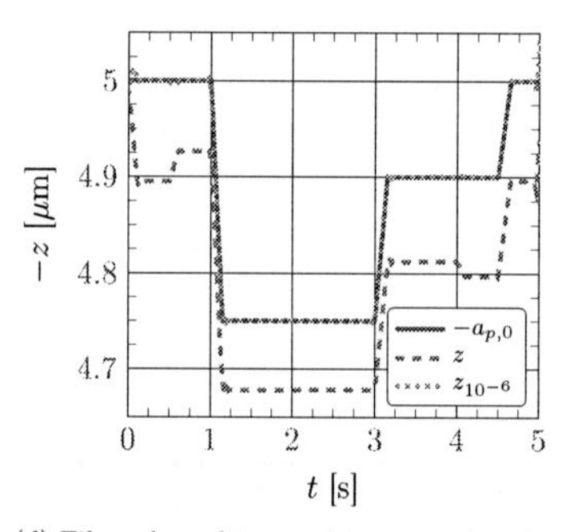

(d) Filtered resulting positions z and z_α from Figure 4.15b.

Figure 4.15: Design problem with machine vibrations.

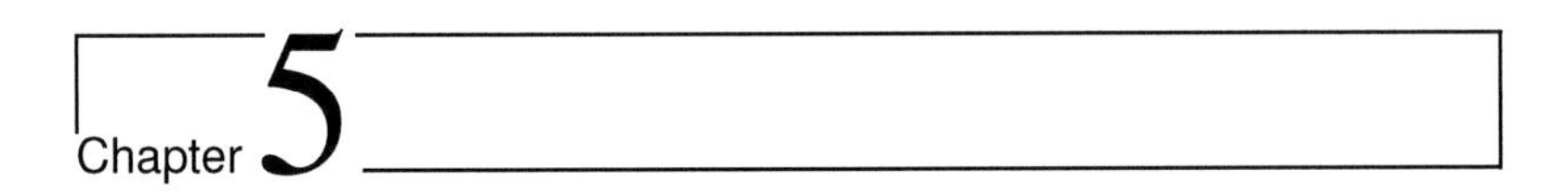

Conclusion

In this thesis, we treated inverse problems appearing in the context of ultra precision turning processes. Because of the strong connection to the application in the field of manufacturing, not only the mathematical theory of inverse problems and regularization is topic of this thesis but also the modeling of ultra precision turning processes.

Therefore, one main part of this thesis concerns the development of a model for ultra precision turning. This model includes in particular the process machine interaction, i.e. the interaction between machine vibrations and the cutting forces occurring during the turning process. On the one hand, the influence of the process is modeled with help of the forces and their resulting moments which they induce to the workpiece and thus to the whole machine structure. The underlying force model for ultra precision turning was thus developed in order to predict the forces in dependence of the relevant process parameters.

On the other hand, the impact of the machine vibrations is included by considering that the vibrating workpiece changes the process parameters. A common approach to incorporate the machine vibrations into the interaction model is to define a new depth of cut as the sum of the input depth of cut and the displacement of the machine in the z-direction, see [12] and the references therein. In our approach we consider the displacements of the workpiece as well as those of the tool caused by deflections and define besides a new depth of cut a new feed velocity as the relative velocity between tool and workpiece. We therefore obtain ordinary differential equations describing these new process parameters.

The machine displacements are determined by solving the linear differential equation for the vibrations. Consequently, the interaction model is described by a coupled system of ordinary differential equations. The coupled system is solved numerically in a time step algorithm, i.e. both equations are solved separately for small time intervals. It would be interesting to investigate the possibility to solve the coupled equation system simultaneously. Therefore, it would be necessary to use an adequate numerical solver because the solution algorithm for the vibration equation could not be further applied. Since the vibration equation do not include damping, on other important topic for future research is the modeling of damping. This is still an open question, although there are some models known in the field of structure dynamics like the Rayleigh damping. Related to this problem is the task of solving the resulting possible nonlinear vibration equation as well as the identification of the damping parameters. Mathematical basis and foundation to this kind of identification problem are given in [31].

The second main part of the thesis is the inversion of the forward operator, i.e. the computation of optimal input parameters for given tool positions. Therefore, the forward operator is formulated as the combination of the parameter-to-state-map, connecting the input parameters with the solution of the system of differential equations, and an observation operator mapping the solution to the tool positions. Instead of solving the nonlinear operator equation, we derived a nonlinear algebraic equation and two linear operator equations which have to be solved in order to identify the distributed parameters.

Since both problems are ill-posed, regularization methods with sparsity constraints are applied which promote sparse solutions. The advantage of such sparse solutions is that they limit the points of machine changes in the machine control. Both inverse problems, the nonlinear as well as the two coupled linear ones, are formulated in Banach spaces, whereas we use for the computation of the minimizer of the Tikhonov-functionals the RFSS-algorithm, which is developed in the Hilbert space setting. Therefore, an interesting problem for future research is to close this theoretical gap. Either to analyze how the problem can be embedded into Hilbert spaces and to guarantee that the computed parameters lie nevertheless in L^∞, or maybe to modify the algorithm such that it can applied in Banach spaces.

We presented two different possible applications, the design problem and the sparse correction problem, where the process parameters can successfully identified with our approach. In order to illustrate our proposed method, we showed various numerical examples, which underline the effectiveness of the approach. The regularization parameter is chosen in all examples by hand, but, in order to simplify the application of the proposed method for users, an appropriate parameter choice rule would be desirable. One lack of this thesis is the missing of real data for testing the inversion process. It would be desirable in future to validate our methods with real experiments. Nevertheless, the numerical examples are promising and we expect our approach to work in real applications.

Appendix A

Useful theorems and facts

For completeness a collection of propositions and theorems are presented in the following which have been used in the previous chapters. We start with some well-known theorems.

Theorem 58 (Implicit function theorem, [64]). *Let X, Y and Z Banach spaces and $G : X \times Y \to Z$ a functional defined on the ball*

$$\mathcal{B}_r = \{x|\, \|x - x_0\|_X \le r, \|y - y_0\|_Y \le r\} .$$

Assume that

1. $G(x_0, y_0) = 0$.
2. *$G(x, y)$ is continuous on $\mathcal{B}_r$.*
3. *$G_y(x, y)$ exits at each $(x, y) \in \mathcal{B}_r$ and is continuous in (x, y).*
4. *The inverse map $[G_y(x_0, y_0)]^{-1} : Z \to Y$ exits.*

Then there exit $\delta > 0$ and $\gamma > 0$ and a continuous mapping $F : X \to Y$, $F(x) = y$ on $\|x - x_0\|_X \le \delta \le r$ such that

1. *$\|F(x) - y\|_Y \le \gamma \le r$ and $F(x_0) = y_0$.*
2. $G(x, F(x)) \equiv 0$.

Proof. The proof can be found in [64]. □

Corollary 59. *Under the hypotheses of the Implicit function theorem, if moreover $G_x(x, y)$ exits on $\mathcal{B}_r$ and is continuous at (x_0, y_0), then $F(x) = y$ has a Fréchet-derivative at x_0 given by*

$$F'(x_0) = -[G_y(x_0, y_0)]^{-1} G_x(x_0, y_0)\, y. \tag{A.1}$$

Proof. We refer to [64] for a proof of this corollary. □

Theorem 60 (Banach Fixed-Point theorem). *Let $M \subseteq X$ be a closed nonempty set in a complete metric space X. Moreover, let $T : M \to M$ a given operator, which maps M into itself. If T is k-contractive, i.e. for fixed k, $0 \le k < 1$, it holds*

$$d(Tx, Ty) \le k\, d(x, y) \quad \text{for all } x, y \in M,$$

then the following assertions are valid.

1. *There is an unique solution of the fixed-point problem*

$$x = Tx, \qquad x \in M\,.$$

2. *The sequence of successive iterations*

$$x_{n+1} = Tx_n \quad n = 0, 1, 2, \ldots$$

with arbitrary $x_0 \in M$ *converges to the solution* x.

Proof. The proof can be found in [66]. □

Proposition 61 (Arzéla-Ascoli)**.** *Let* E *be a Banach space and* K *be a compact metric space. Then the subset* $M \subseteq C(K, E)$ *is relatively compact (precompact) if the following two conditions are valid:*

1. M *is equicontinuous, i.e. for all* $t \in K$ *and for all* $\varepsilon > 0$ *there exits a open neighborhood* $\mathcal{U}$ *of* t *in* K *such that*

$$\|f(t) - f(s)\|_E < \epsilon \quad \textit{for all } s \in \mathcal{U} \textit{ and all } f \in M\,.$$

2. *The set* $M(t) := \{f(t) \,|\, f \in M\}$ *is relatively compact in* E *for all* $t \in K$.

In the case that E *is finite dimensional,* M *is precompact if* M *is equicontinuous and bounded.*

Proof. We refer to [63], Theorem II.3.4 for a proof. □

In the following some facts are summarized which are needed in Subsection 2.3.

Proposition 62 ([26], Theorem 1.4.3)**.** *Let* $(\Omega, \mathcal{A}, \mu)$ *be a measure space,* $\mathcal{I}$ *a measurable subset of* Ω *and* D *a subset of a Banach space* E*. Assume that* $f : \mathcal{I} \times D \to E$ *is a Carathéodory function fullfilling the properties* (F1) *and* (F2)*, i.e.*

1. $f(\cdot, \xi)$ *is* μ*-measurable for each* $\xi \in D$.

2. $f(t, \cdot)$ *is continuous in* D *for a.a.* $t \in \mathcal{I}$.

If moreover $u : \mathcal{I} \to D$ *is* μ*-measurable, then* $f(\cdot, u(\cdot))$ *is* μ*-measurable.*

Proof. Suppose that $u : \mathcal{I} \to D$ is μ-measurable and let $\{u_n\}_n$ be a sequence of step functions which converges pointwise a.e. to u on $\mathcal{I}$. Then by the first property follows that $f(\cdot, u_n(\cdot))$ is μ-measurable for all $n \in \mathbb{N}$. The second property forces that

$$\lim_{n\to\infty} f(t, u_n(t)) = f(t, u(t)) \quad \text{for a.a. } t \in \mathcal{I}\,.$$

Consequently, we have shown that $f(\cdot, u(\cdot))$ is μ-measurable. □

Lemma 63. *Let* $\mathcal{I}$ *be an interval and* P *a metric space. Assume that there is a sequence* $\{p_n\}_n$ *and a function* p *in* $L^\infty(\mathcal{I}, P)$ *such that* $p_n \to p$ *in* $L^\infty(\mathcal{I}, P)$ *as* $n \to \infty$*. Then the set of* p_n *and* p *is equibounded, i.e. there is a compact subset* $K \subseteq P$ *so that* $p(t) \in K$ *and* $p_n(t) \in K$ *for almost all* $t \in \mathcal{I}$ *and all* n.

Proof. See [55], Remark C.1.3. □

Let $f : \mathcal{U} \subseteq X \to Y$ differentiable in every point $x \in \mathcal{U}$, then $f' : \mathcal{U} \to L(X,Y)$. If f' is again differentiable in $x \in \mathcal{U}$, then $f''(x) \in L(X, L(X,Y))$ is called second Fréchet-derivative at x. The space $L(X, L(X,Y))$ can be identified with the space $L_2(X^2, Y)$ of bilinear continuous maps. Consequently the map f is twice Fréchet-differentiable at x, if there exits a bilinear continuous operator $f''(x) \in L_2(X^2, Y)$ such that

$$F'(x+k)h - F'(x)h = F''(x)(h,k) + R(x,h,k) \qquad (h, k \in X)$$

with

$$\lim_{\|k\| \to 0} \frac{\|R(x,h,k)\|}{\|k\|} = 0 \qquad \text{for } \|h\| = 1 .$$

In the same way higher derivatives can be defined. Thus, we can now formulate the Taylor's Theorem.

Theorem 64 (Generalized Taylor's Theorem). *Let X and Y Banach-spaces, $\mathcal{U}(x) \subseteq X$ an open convex neighborhood of x and $f : \mathcal{U} \to Y$. If the F-derivatives $f^{(n)}$ actually exit on $\mathcal{U}(x)$, then f can be written in the expansion*

$$f(x+h) = f(x) + \sum_{k=1}^{n-1} \frac{1}{k!} f^{(k)}(x)\, h^k + R_n \tag{A.2}$$

with remainder bound

$$\|R_n\| = \frac{1}{n!} \sup_{0<t<1} \left\| f^{(n)}(x + t\,h)\, h^n \right\| . \tag{A.3}$$

If moreover $f^{(n)}$ is continuous on $\mathcal{U}(x)$, then

$$R_n = \int_0^1 \frac{(1-t)^{n-1}}{(n-1)!}\, f^{(n)}\,(x + t\,h)\; h^n \, dt . \tag{A.4}$$

More about differential equations

B.1 Linear Differential Equations

Let E be a finite dimensional Banach space over $\mathbb{K}$ and $\mathcal{I}$ an open interval in $\mathbb{R}$. We will investigate the linear nonhomogeneous differential equation

$$\dot{u}(t) = A(t)u(t) + b(t) \qquad u(t_0) = u_0 \tag{B.1}$$

with $A \in C(\mathcal{I}, L(E))$ and $b \in C(\mathcal{I}, E)$. Theorem 21 states that there exits for each $(t_0, u_0) \in \mathcal{I} \times E$ a unique solution $u : \mathcal{I} \to E$. Moreover, we know that for

$$\phi : \mathcal{I} \times \mathcal{I} \times E \to E \qquad (t, t_0, u_0) \mapsto u(t)$$

it holds that $\phi \in C(\mathcal{I} \times \mathcal{I} \times E, E)$. From this the next theorem follows immediately.

Theorem 65 ([2], Theorem 11.2). *The set of all solutions of the homogeneous equation*

$$\dot{u}(t) = A(t)u(t) \tag{B.2}$$

forms a vector subspace V of $C^1(\mathcal{I}, E)$ of dimension $n = \dim(E)$. For each fixed $t_0 \in \mathcal{I}$ the mapping $\xi \mapsto \phi(\cdot, t_0, \xi)$ defines an isomorphism from E to V.

Proof. It follows from the unique solvability that for $\lambda, \mu \in \mathbb{K}$ and $\xi, \eta \in E$ each side of

$$\phi(\cdot, t_0, \lambda\xi + \nu\eta) = \lambda\phi(\cdot, t_0, \xi) + \mu\phi(\cdot, t_0, \eta)$$

is a solution of (B.2) with initial condition $u(t_0) = \lambda\xi + \mu\eta$. Thus, the mapping $\xi \mapsto \phi(\cdot, t_0, \xi)$ is linear. From $\phi(\cdot, t_0, \xi) = 0$ follows that $\xi = 0$. Therefore the mapping $\xi \mapsto \phi(\cdot, t_0, \xi)$ is injective, hence a vector space isomorphism from E to V. □

We call every solution of

$$\dot{X}(t) = A(t)X(t) \text{ in } L(E) \tag{B.3}$$

a *solution matrix* of the differential equation

$$\dot{x}(t) = A(t)x(t) \text{ in } E. \tag{B.4}$$

Theorem 66. *Let $v \in C^1(\mathcal{I}, E)$ an arbitrary solution of* (B.2) *and V be the solution space of the corresponding homogenous equation* (B.1). *Then the set of all solutions of the inhomogeneous equation* (B.1) *forms an affine subspace $v + V$ of $C^1(\mathcal{I}, E)$.*

Proof. If there are two arbitrary solutions u and v of (B.1), then clearly $u - v = A(t)(u - v) \in V$. □

Remark 67. We denote the space of all $(n \times n)-$matrices over $\mathbb{K}$ with $\mathbb{M}^n(\mathbb{K})$. If $E = \mathbb{K}^n$, we can identify $A(t)$ with its matrix representation with respect to the canonical basis of $\mathbb{K}^n$, $A(t) = [a_{i,j}]_{1 \leq i,j, \leq n}$. If $\{x_i, i = 1, \dots, n\}$ is a fundamental set of $\dot{x} = A(t)x$, the matrix $X(t) = [x_1(t), \dots, x_n(t)]$ is called fundamental matrix. If additionally $X(t_0) = \mathbb{I}_{\mathbb{K}^n} =: \mathbb{I}_n$, then $X_{t_0} = X$ is called special fundamental matrix at time t_0. It is the unique solution of

$$\dot{X}(t) = A(t)X(t), \qquad X(t_0) = \mathbb{I}_n$$

in the space of all $(n \times n)-$matrices over $\mathbb{K}$. It follows that for each $\xi \in \mathbb{K}^n$ that the unique solution of

$$\dot{u}(t) = A(t)u(t), \qquad u(t_0) = \xi,$$

is given by

$$u(t) = X_{t_0}(t)\xi.$$

Theorem 68 (Variation of Constants Formula, Theorem 11.13 in [2]). *Let $t_0 \in \mathcal{I}$ and $\xi \in E$. Then the linear IVP*

$$\dot{u}(t) = A(t)u(t) + b(t), \qquad u(t_0) = \xi\,, \tag{B.5}$$

has an unique solution $u(\cdot) = \phi(\cdot, t_0, \xi)$, which is given by

$$u(t) = \Phi(t, t_0)\,\xi + \int_{t_0}^{t} \Phi(t, s)\, b(s)\, ds, \quad t \in \mathcal{I}\,. \tag{B.6}$$

Hereby, $\Phi(\cdot, \tau)$ called evolution operator is the global solution of the IVP

$$\dot{Y}(t) = A(t)Y(t), \qquad Y(\tau) = \mathbb{I}_E\,, \tag{B.7}$$

in $L(E)$ for all $\tau \in \mathcal{I}$. Moreover, $\Phi \in C^1(\mathcal{I} \times \mathcal{I}, L(E))$ and

1. *$\Phi(\tau, \tau) = \mathbb{I}_E, \qquad$ for all $\tau \in \mathcal{I}$,*
2. *$\Phi(t, \tau)\,\Phi(\tau, s) = \Phi(t, s)$ for all $\tau, t, s \in \mathcal{I}$.*

Proof. Let X be an arbitrary fundamental matrix of the equation $\dot{Y}(t) = A(t)Y(t)$. We define the evolution operator as

$$\Phi(t, s) := X(t)X^{-1}(s), \qquad t, s \in \mathcal{I}. \tag{B.8}$$

Clearly, $\Phi \in C^1(\mathcal{I} \times \mathcal{I}, L(E))$, and by differentiation we obtain

$$D_1\Phi(t, s) = \dot{X}(t)X^{-1}(s) = A(t)X(t)X^{-1}(s) = A(t)\Phi(t, s)\,. \tag{B.9}$$

Moreover, $\Phi(t, t) = X(t)X^{-1}(t) = \mathbb{I}_E$. Therefor, $\Phi(\cdot, \tau)$ is the unique solution of the IVP of

$$\dot{Y}(t) = A(t)Y(t)\,, \qquad Y(\tau) = \mathbb{I}_E\,,$$

in $L(E)$ for every $\tau \in \mathcal{I}$, i.e. $\Phi(\cdot, \tau)$ is the special fundamental matrix of $\dot{u}(t) = A(t)u(t)$ at time τ :

$$\Phi(t, \tau) = X_\tau(t), \qquad \text{for all } t, \tau \in \mathcal{I}. \tag{B.10}$$

By (B.8) we deduce for $\tau, t, s, \in \mathcal{I}$ that

$$\Phi(t, \tau)\Phi(\tau, s) = X(t)X^{-1}(\tau)X(\tau)X^{-1}(s) = X(t)X^{-1}(s) = \Phi(t, s)$$

and hence,

$$\Phi(t, \tau)\Phi(\tau, t) = \Phi(t, t) = \mathbb{I}_E.$$

It follows that

$$\Phi(t, \tau) = [\Phi(\tau, t)]^{-1}, \qquad \text{for all } t, \tau \in \mathcal{I}. \tag{B.11}$$

Comparing with (B.10), we see that $X_\tau(t)^{-1} = \Phi(\tau, t)$, and hence we obtain

$$X_\tau(t)X_\tau^{-1}(s) = \Phi(t, \tau)\Phi(\tau, s) = \Phi(t, s).$$

We apply now a linear transformation to the nonhomogeneous equation $\dot{u} = Au + b$, i.e. we set $u(t) = \Phi(t, s)y(t)$. We obtain for the transformed equation

$$\begin{aligned}
\dot{y}(t) &= \frac{\mathrm{d}}{\mathrm{d}t}\left(\Phi^{-1}(t, s)u(t)\right) = \dot{\Phi}^{-1}(t, s)u(t) + \Phi^{-1}(t, s)\dot{u}(t), \\
&= \dot{\Phi}^{-1}(t, s)\Phi(t, s)y(t) + \Phi^{-1}(t, s)\left[A(t)\Phi(t, s)y(t) + b(t)\right], \\
&= \dot{\Phi}^{-1}(t, s)\Phi(t, s)y(t) + \Phi(s, t)A(t)\Phi(t, s)y(t) + \Phi(s, t)b(t),
\end{aligned} \tag{B.12}$$

where we used (B.11) in the last line. By $\Phi(t, s)\Phi^{-1}(t, s) = \mathbb{I}_n$, we derive $\dot{\Phi}(t, s)\Phi^{-1}(t, s) + \Phi(t, s)\dot{\Phi}^{-1}(t, s) = 0$, and therefore

$$\dot{\Phi}^{-1}(t, s) = -\Phi^{-1}(t, s)\dot{\Phi}(t, s)\Phi^{-1}(t, s) = -\Phi(s, t)\dot{\Phi}(t, s)\Phi(s, t).$$

Thus, the transformed equation (B.12) becomes

$$\begin{aligned}
\dot{y}(t) &= \Phi(s, t)\left[-\dot{\Phi}(t, s)\Phi(s, t)\Phi(t, s) + A(t)\Phi(t, s)\right]y(t) + \Phi(s, t)b(t), \\
&= \Phi(s, t)\left[-\dot{\Phi}(t, s) + A(t)\Phi(t, s)\right]y(t) + \Phi(s, t)b(t),
\end{aligned}$$

which takes the simple form

$$\dot{y}(t) = \Phi(s, t)b(t),$$

because of (B.9) (i.e. because Φ is the fundamental matrix). A solution of this equation is the function

$$y(t) = \int_{t_0}^{t} \Phi(s, \tau)b(\tau)\,\mathrm{d}\tau,$$

and thus

$$u(t) = \Phi(t, s)\int_{t_0}^{t} \Phi(s, \tau)b(\tau)\,\mathrm{d}\tau = \int_{t_0}^{t} \Phi(t, \tau)b(\tau)\,\mathrm{d}\tau$$

is a special solution of the nonhomogeneous equation (B.5). By Theorem 66 we therefore get that all solutions of (B.5) are given by

$$u(t) = \Phi(t, t_0)\xi + \int_{t_0}^{t} \Phi(t, \tau)b(\tau)\,\mathrm{d}\tau,$$

what completes the statement of the Theorem. □

Nomenclature

A^*	Adjoint operator of A
$A^\dagger$	Pseudo-inverse of A
α	Regularization parameter
A_c	Cross sectional area of cut
AC	Space of absolutely continuous functions
a_p	Depth of cut
$a_{p,\alpha}$	Computed input depth of cut for a regularization parameter α
$a_p^0(t)$	Input depth of cut for the forwad model
$\mathcal{B}_\rho(\xi_0)$	Ball with radius ρ centered at ξ_0
β_i	Angular displacements of the workpiece around the axis in direction i, $i = x, y, z$
$C(\mathcal{I}, E)$	Space of continuous functions $f : \mathcal{I} \to E$
$C^k(\mathcal{I}, E)$	Subspace of $C^k(\mathcal{I}, E)$ of k-times continuous differentiable functions $f : \mathcal{I} \to E$
$\mathcal{D}(f)$	Domain of the function f
$D_1 f(x, y)$	Partial derivative of $f(x, y)$ with respect to x
$\partial f\,(x_0)$	Subdifferential of f in x_0
Δ_i	Displacements of the workpiece in direction i, $i = x, y, z$
δ_i	Deflection of the tool in direction i, $i = x, y, z$
E	Banach space
F	Forward operator
F_c	Cutting force
f	Feed rate

F_f	Feed force
F_p	Thurst or passive force
$f_x(x,y)$	Partial derivative of $f(x,y)$ with respect to x
Im	Imaginary part of a complex number
$\mathcal{I} \subseteq \mathbb{R}$	Interval in $\mathbb{R}$
$K(X,Y)$	The set of compact linear operators $A \in L(X,Y)$
k_c	Specific cutting force
$L(X,Y)$	Space of linear and continuous maps from X to Y
$L^1(X)$	Banach space of all integrable functions
l^p	Sequence space
$L^p(X)$	Banach space of all p -integrable functions
$L^\infty(\mathcal{I}, P)$	Space of measurable and essentially bounded functions
$L^\infty_{loc}(\mathcal{I}, P)$	Space of measurable and locally essentially bounded functions
$\boldsymbol{\lambda}$	Vector of displacements of the machine model
$\mathbf{M}$	Mass matrix
n	Rotational speed
$\Omega(x)$	Penality term for the Tikhonov-functional
Φ	Evolution operator of the homogenous differential equation
φ	Parameter-to-state-map
$\mathcal{R}(A)$	Range of A
Re	Real part of a complex number
r	Radius of the workpiece
r_ϵ	Tool nose radius
$\mathbf{S}$	Stiffness matrix
$\mathrm{Sgn}(x)$	Set-valued Sign function
v_c	Cutting velocity
v_f	Feed velocity $v_f = n\,f$
$v_{f,\alpha}$	Computed input feed velocity for a regularization parameter α
$v_f^0(t)$	Input feed velocity for the forwad model

v_x	Actual feed velocity
$W^{1,p}(\mathcal{I}, E)$	Sobolev space of p-integrable functions $f : \mathcal{I} \to E$ with p-integrable first derivative
x^*	Desired x-posistion
x_α	Computed x-position for a regularization parameter α
z^*	Desired z-posistion
z_α	Computed z-position for a regularization parameter α

Bibliography

[1] Y. Altintas. *Manufacturing Automation: Metal Cutting Mechanics, Machine Tool Vibrations, and CNC Design.* Cambridge Univ. Press, Cambridge [u.a.], 2000.

[2] H. Amann. *Ordinary Differential Equations - An Introduction to Nonlinear Analysis.* De Gruyter, 1990.

[3] M. Andres, H. Blum, C. Brandt, C. Carstensen, P. Maass, J. Niebsch, R. Rademacher, R. Ramlau, A. Schröder, E.-P. Stephan, and S. Wiedemann. *Process Machine Interactions*, chapter Adaptive finite elements and mathematical optimization methods. Lectures Notes in Production Engineering. Springer, 2013.

[4] A. B. Bakushinskii. Remarks on choosing a regularization parameter using the quasi-optimality and the ratio criterion. In *USSR Comput. Math. Math. Phys.*, 24(8):181–182, 1984.

[5] F. Bauer and M. A. Lukas. Comparing parameter choice methods for regularization of ill-posed problems. In *Mathematics and Computers in Simulation*, 81(9):1795 – 1841, 2011.

[6] C. Brandt, A. Krause, E. Brinksmeier, and P. Maaß. Force modelling in diamond machining with regard to the surface generation process. In *Proceedings of the 9th International Conference and Exhibition on Laser Metrology, machine tool, CMM and robotic performance, LAMDAMAP 2009, London, 30.06.-02.07.2009*, 377–386. 2009.

[7] C. Brandt, A. Krause, J. Niebsch, J. Vehmeyer, E. Brinksmeier, P. Maass, and O. Riemer. *Process Machine Interactions*, chapter Surface Generation Process with Consideration ot the Balancing State in Diamond Machining. Lectures Notes in Production Engineering. Springer, 2013.

[8] C. Brandt, P. Maass, I. Piotrowska-Kurczewski, S. Schiffler, O. Riemer, and E. Brinksmeier. Mathematical methods for optimizing high precision cutting operations. In *International Journal of Nanomanufacturing*, 8(4):306–325, 2012.

[9] C. Brandt, J. Niebsch, P. Maaß, and R. Ramlau. Simulation of process machine interaction for ultra precision turning. In *Proceedings of the 2nd International*

Conference on Process Machine Interactions, June 10-11, 2010, Vancouver, Canada. 2010.

[10] C. Brandt, J. Niebsch, R. Ramlau, and P. Maass. Modeling the influence of unbalances for ultra-precision cutting processes. In *ZAMM - Journal of Applied Mathematics and Mechanics / Zeitschrift für Angewandte Mathematik und Mechanik*, 91(10):795–808, 2011.

[11] C. Brandt, J. Niebsch, and J. Vehmeyer. Modelling of ultra-precision turning process in consideration of unbalances. In *Advanced Materials Research*, 223:839–848, 2011.

[12] C. Brecher, M. Esser, and S. Witt. Interaction of manufacturing process and machine tool. In *CIRP Annals - Manufacturing Technology*, 58(2):588–607, 2009.

[13] K. Bredies and D. A. Lorenz. Iterated hard shrinkage for minimization problems with sparsity constraints. In *SIAM Journal on Scientific Computing*, 30(2):657–683, 2008.

[14] K. Bredies and D. A. Lorenz. Linear convergence of iterative soft-thresholding. In *Journal of Fourier Analysis and Applications*, 14(5–6):813–837, 2008.

[15] E. Brinksmeier, O. Riemer, R. Gläbe, B. Lünemann, C. Kopylow, C. Dankwart, and A. Meier. Submicron functional surfaces generated by diamond machining. In *CIRP Annals - Manufacturing Technology*, 59(1):535 – 538, 2010.

[16] I. Cioranescu. *Geometry of Banach spaces, duality mappings and nonlinear problems.* Mathematics and its applications ; 62. Kluwer, Dordrecht, 1990.

[17] I. Daubechies, M. Defrise, and C. De Mol. An iterative thresholding algorithm for linear inverse problems with a sparsity constraint. In *Communications in Pure and Applied Mathematics*, 57(11):1413–1457, 2004.

[18] K. Deimling. *Ordinary Differential Equations in Banach Spaces.* Lecture Notes in Mathematics ; 596. Springer, Berlin [u.a.], 1977.

[19] H. W. Engl, M. Hanke, and A. Neubauer. *Regularization of Inverse Problems*, volume 375 of *Mathematics and its Applications.* Kluwer Academic Publishers Group, Dordrecht, 2000.

[20] M. Eynian. *Chatter Stability of General Turning Operations With Process Damping.* Ph.D. thesis, The Unversity of British Columbia, 2010.

[21] M. Eynian and Y. Altintas. Chatter stability of general turning operations with process damping. In *Journal of Manufacturing Science and Engineering*, 131(4):041005, 2009.

[22] A. F. Filippov. *Differential equations with discontinuous righthand sides.* Mathematics and its applications, Soviet series 18. Kluwer, Dordrecht [u.a.], 1988.

[23] R. Gasch and K. Knothe. *Strukturdynamik Bd. 2: Kontinua und ihre Diskretisierung.* Springer, Berlin, 1989.

[24] R. Griesse and D. A. Lorenz. A semismooth Newton method for Tikhonov functionals with sparsity constraints. In *Inverse Problems*, 24(3):035007 (19pp), 2008.

[25] J. Hadamard. *Lectures on Cauchy's Problem in Linear Partial Differential Equations.* Yale University Press, 1923.

[26] S. Heikkilä and V. Lakshmikantham. *Monotone iterative techniques for discontinuous nonlinear differential equations.* Monographs and textbooks in pure and applied mathematics 181. Dekker, New York, 1994.

[27] H. Heuser. *Gewöhnliche Differentialgleichungen: Einführung in Lehre und Gebrauch.* Mathematische Leitfäden. Teubner, Stuttgart, 1989.

[28] B. Jin, D. A. Lorenz, and S. Schiffler. Elastic-net regularization: Error estimates and active set methods. In *Inverse Problems*, 25(11):115022 (26pp), 2009.

[29] X. Jin and Y. Altintas. Slip-line field model of micro-cutting process with round tool edge effect. In *Journal of Materials Processing Technology*, 211:339–355, 2011.

[30] F. Jones. *Lebesgue integration on Euclidean space.* Jones and Bartlett books in mathematics. Jones and Bartlett, Boston [u.a.], 1993.

[31] B. Kanning. *Instationary vibrational analysis for impulse-type stimulated structures.* Ph.D. thesis, University of Bremen, 2012.

[32] M. Kaymakci, Z. Kilic, and Y. Altintas. Unified cutting force model for turning, boring, drilling and milling operations. In *International Journal of Machine Tools and Manufacture*, 54 - 55(0):34 – 45, 2012.

[33] K. S. Kazimierski. *Aspects of regularization in Banach spaces.* Ph.D. thesis, University of Bremen, Berlin, 2011.

[34] J. Köhler. *Berechnung der Zerspankräfte bei variierenden Spanungsquerschnittsformen.* Ph.D. thesis, Leibniz Universität Hannover, 2010.

[35] O. Kienzle. Die Bestimmung von Kräften und Leistungen an spanenden Werkzeugen und Werkzeugmaschinen. In *Z. VDI*, 94(11):299–305, 1952.

[36] A. Krause and E. Brinksmeier. Process forces in diamond machining with consideration of unbalances. In *Proceedings of the 2nd International Conference on Process Machine Interactions, June 10-11, 2010, Vancouver, Canada.* 2010.

[37] C. Kravaris and J. Steinfeld. Identification of parameters in distributed parameter systems by regularization. In *SIAM Journal on Control and Optimization*, 23(2):217–241, 1985.

[38] H. Lee, A. Battle, R. Raina, and A. Y. Ng. Efficient sparse coding algorithms. In B. Schölkopf, J. Platt, and T. Hoffman (editors), *Advances in Neural Information Processing Systems 19*, 801–808. MIT Press, Cambridge, MA, 2007.

[39] W. Lee and C. Cheung. A dynamic surface topography model for the prediction of nano-surface generation in ultra-precision machining. In *International Journal of Mechanical Sciences*, 43(4):961 – 991, 2001.

[40] A. K. Louis. *Inverse und schlecht gestellte Probleme.* Teubner-Studienbücher, Mathematik. Teubner, Stuttgart, 1989.

[41] M. Malekian, S. Park, and M. Jun. Investigation of critical chip thickness and micro ploughing forces. In *Proceedings of the 2nd International Conference on Process Machine Interactions, June 10-11, 2010, Vancouver, Canada.* 2010.

[42] J. Niebsch and R. Ramlau. Imbalances in high precision cutting machinery. In *Proceedings of the ASME 2009 IDETC/CIE. San Diego, 2009, DETC/VIB-86937.* 2009.

[43] J. Niebsch, R. Ramlau, and C. Brandt. On the interaction of unbalances and surface quality in ultra-precision cutting machinery. In *SIRM 2011, Darmstadt, Germany.* 2011.

[44] R. Palm. *Numerical comparison of regularization algorithms for solving ill-posed problems.* Ph.D. thesis, University of Tartu, Estonia, 2010.

[45] I. Piotrowska, C. Brandt, H. R. Karimi, and P. Maass. Mathematical model of micro turning process. In *The International Journal of Advanced Manufacturing Technology*, Volume 45(1):33–40, 2009.

[46] I. Piotrowska-Kurczewski and J. Vehmeyer. Simulation model for micro-milling operations and surface generation. In *Advanced Materials Research*, 223:849–858, 2011.

[47] R. A. Ressel. *A parameter identification problem involving a nonlinear parabolic differential equation.* Ph.D. thesis, University of Bremen, 2012.

[48] A. Rieder. *Keine Probleme mit inversen Problemen: eine Einführung in ihre stabile Lösung.* Vieweg, Wiesbaden, 1 edition, 2003.

[49] O. Riemer. *Trennmechanismen und Oberflächenfeingestalt bei der Mikrozerspanung kristalliner und amorpher Werkstoffe.* Ph.D. thesis, Universität Bremen, 2001.

[50] O. Riemer. Advances in ultra precision manufacturing. In *2011 JSPE Spring Meeting.* 2011.

[51] R. T. Rockafellar and R. J.-B. Wets. *Variational analysis.* Grundlehren der mathematischen Wissenschaften 317. Springer, Berlin, 1998.

[52] O. Rott, D. Hömberg, and C. Mense. A comparison of analytical cutting force models. In *WIAS Preprint 1151.*

[53] O. Scherzer, M. Grasmair, H. Grossauer, M. Haltmaier, and F. Lenzen. *Variational methods in imaging.* Springer, 2009.

[54] S. Schiffler. *The elastic net: Stability for sparsity methods.* Ph.D. thesis, University of Bremen, 2010.

[55] E. D. Sontag. *Mathematical control theory: deterministic finite dimensional systems.* Texts in applied mathematics 6. Springer, New York, NY [u.a.], 2 edition, 1998.

[56] N. Taniguchi. Current status in, and future trends of, ultraprecision machining and ultrafine materials processing. In *CIRP Annals - Manufacturing Technology*, 32(2):573 – 582, 1983.

[57] A. N. Tikhonov. Regularization of incorrectly posed problems. In *Soviet Mathematics Doklady*, 4:1624–1627, 1963.

[58] A. N. Tikhonov and V. B. Glasko. Use of the regularization method in nonlinear problems. In *USSR Comput. Math. Math. Phys.*, 5(3):93–107, 1965.

[59] H. K. Tönshoff and B. Denkena. *Spanen: Grundlagen.* Springer, Berlin, 2. edition, 2004.

[60] J. Vehmeyer, I. Piotrowska-Kurczewski, and S. Twardy. A surface generation model for micro cutting processes with geometrically defined cutting edges. In *37th International MATADOR Conference.* 2012.

[61] F. Vollertsen, D. Biermann, H. Hansen, I. Jawahir, and K. Kuzman. Size effects in manufacturing of metallic components. In *CIRP Annals - Manufacturing Technology*, 58(2):566 – 587, 2009.

[62] M. Weber, H. Autenrieth, J. Kotschenreuther, P. Gumbsch, V. Schulze, D. Lohe, and J. Fleischer. Influence of friction and process parameters on the specific cutting force and surface characteristics in micro cutting. In *Machining Science and Technology*, 12(4):474–497, 2008.

[63] D. Werner. *Funktionale und Operatoren.* Springer-Lehrbuch. Springer Berlin Heidelberg, 2007.

[64] A. Wouk. *A course of applied functional analysis.* Pure and applied mathematics. Wiley, New York [u.a.], 1979.

[65] E. Zeidler. *Nonlinear functional analysis and its applications III: Variational Methods and Optimization.* Springer, New York, NY [u.a.], 1985.

[66] E. Zeidler. *Nonlinear functional analysis and its applications I: Fixed-point theorems.* Springer, New York, NY [u.a.], 1986.

[67] S. Zhou and J. Shi. Active balancing and vibration control of rotating machinery: A survey. In *The Shock and Vibration Digest*, 33(4):361–371, 2001.